PATHOGEN REMOVAL MECHANISMS IN MACROPHYTE AND ALGAL WASTE STABILIZATION PONDS

Pathogen Removal Mechanisms in Macrophyte and Algal Waste Stabilization Ponds

DISSERTATION

Submitted in fulfilment of the requirements of
the Academic Board of Wageningen University and
the Academic Board of the UNESCO-IHE Institute for Water Education
for the Degree of DOCTOR
to be defended in public
on Wednesday, 29 November 2006 at 15.30 hours
in Delft, The Netherlands

by

ESI AWUAH
born in Akim-Oda, Ghana

Promoter:	Prof. dr. H.J. Gijzen, Professor of Environmental Biotechnology UNESCO-IHE Institute of Water Education, The Netherlands
Co-promoter:	Dr. H.J. Lubberding. Senior lecturer Microbiology UNESCO-IHE Institute of Water Education, The Netherlands
Awarding Committee	Prof. dr. G.L. Amy UNESCO-IHE Institute of Water Education, The Netherlands
	Prof. K.A. Andam Kwame Nkrumah University of Science and Technology, Kumasi, Ghana
	Prof. dr. ir. A.J.M Stams Wageningen University Research Centre, The Netherlands
	Dr. ir. N.P. van der Steen UNESCO-IHE Institute of Water Education, The Netherlands
	Prof. dr. ir. W. Verstraete Ghent University, Belgium

Taylor & Francis is an imprint of the Taylor & Francis Group, an informa business

© 2006, Esi Awuah

Published by:
Taylor & Francis/Balkema
PO Box 447, 2300 AK Leiden, The Netherlands
e-mail: Pub.NL@tandf.co.uk
www.balkema.nl, www.taylorandfrancis.co.uk, www.crcpress.com

ISBN10 0-415-41695-7 (Taylor & Francis Group)
ISBN13 978-0-415-41695-5 (Taylor & Francis Group)
ISBN 90-8504-500-2 (Wageningen University)

To God be the Glory

Dedicated to

My husband, Richard Tuyee Awuah and my sons, Ato and Kobbina

"For those God foreknew, He also predestined to be conformed to the likeness of His Son that He might be the First born among many brothers. Those He predestined, He also called; those He called, He also justified; those He justified, He also glorified. Romans 8:29-30.
NIV

ACKNOWLEDGEMENTS

I wish to thank God Almighty who granted me the Divine Grace to pursue this research. I give Him my heartfelt gratitude for the strength especially when I was weak. The divine wisdom and favours received from God are too numerous to count. 'Nyame W'aseda ye Bebre'.

I want to thank the people of the Netherlands from the smallest to the greatest through the SAIL Foundation for financially supporting this PhD program. I am grateful to my Promoter Prof. Huub J. Gijzen for modifying the topic of my PhD research and carefully supervising the program, for his encouragement and moral support. I am also grateful to my mentor Dr. Henk Lubberding, who painstakingly, edited my manuscripts with very good constructive criticisms. I am also grateful to Prof. Emmanuel Frempong for his encouragement and good wishes for my work.

I want to thank all my Colleagues; Omar Zimmo, Julia Rosa, Moussa, Okorut, Richard Buamah, Kwabena Nyarko and all the PhD students from 1998 to 2006 for their support. I am also grateful to the many beautiful faces at UNESCO-IHE, which assured me of love and comfort, making my coming to Delft always a joyful one.

I would also like to thank God for Dr. S. N. Odai, Dr. Anyemedu, and Mr. Trifunovic, Prof. Kwesi Andam, the Vice Chancellor of KNUST and the wife Prof. Mrs. Aba Andam for their encouragement, moral and spiritual support. I am also grateful to the Technicians of the Environmental Quality Engineering Laboratory KNUST (Mr. Bruce, Mr. Parker, Mr. Botchway and Mr. Antwi) for their immense assistance during the laboratory analyses.

I am highly indebted to all the students who worked with me to produce this PhD thesis. I would also like to extend my profound gratitude to all the young maidens who took care of my house whenever I was out of the country; Vivian, Charity, Mercy and Florence. I am also thankful to members of my Church, Mount Zion Methodist in Kotei, Ghana and Mount Zion Redeemed Christian Church of God, Delft, Netherlands who supported me spiritually with their prayers and warmth at fellowship, gifts etc. I want to thank my parents, Mr. Emmanuel Broni Asare, and my mother Margaret Annan for their prayers and support. I am especially grateful to my mother for providing most of the hand gloves I needed and all the aluminium foil I used in Ghana for my research work. I am most thankful for the prayer support and good wishes of my family members and friends.

My children Ato and Kobbina suffered a lot in my absence. I dedicate this thesis to them.

Finally, I would like to express my most profound gratitude to my husband Professor Richard Tuyee Awuah for encouraging me to continue my education, his prayers and good wishes and for the numerous phone calls, e-mail and text messages to assure me of his love and support.

May God remember the people I might have forgotten and reward them for all the help they gave me.

LIST OF ABBREVIATIONS

BOD	Biochemical Oxygen Demand
CO_3^{-2}	Carbonate ion
COD	Chemical Oxygen Demand
CSTR	Completely Stirred Tank Reactor
DO	Dissolved Oxygen (mg/L)
FSAU	Food Security Analysis Unit
IHE	International Institute for Infrastructural, Hydraulic and Environmental Engineering (can be removed)
HCO_3^-	Bicarbonate ion
KNUST	Kwame Nkrumah University of Science and Technology
NH_4^+	Ammonium ion
NIV	New International Version
SAIL	Netherlands Co-operation Fund
SCF	Save the Children Foundation
SUNY	State University of New York
TDS	Total Dissolved Solids
TSS	Total Suspended Solids
UASB	Upflow Anaerobic Sludge Blanket Reactor
UNESCO	United Nations Educational, Scientific and Cultural Organization
USEPA	United States Environmental Protection Agency
WSESP	Water Supply and Environmental Sanitation Project
WHO	World Health Organization

CONTENTS

ACKNOWLEDGEMENTS _____ viii

LIST OF ABBREVIATIONS _____ ix

CONTENTS _____ xi

Chapter One _____ 1

Introduction _____ 1

Chapter Two _____ 29

Environmental Conditions and Enterococci Removal in Macrophyte and Algal-Based
Domestic Wastewater Treatment Systems _____ 29

Chapter Three _____ 43

The Effect of pH on Enterococci Removal in Water Lettuce, Duckweed and Algal
Ponds _____ 43

Chapter Four _____ 55

Environmental Conditions and Effect of pH on Faecal Coliform Removal in
Macrophyte and Algal Ponds _____ 55

Chapter Five _____ 69

Effect of pH Fluctuations on Pathogenic Bacteria Removal in Domestic Wastewater 69

Chapter Six _____ 79

Comparative Performance studies of macrophyte and algal pondsUsing Low Strength
Sewage _____ 79

Chapter Seven _____ 93

Comparative Performance Studies of Macrophyte and Algal Ponds Using Medium
Strength Sewage _____ 93

Chapter Eight _____ 113

The Role of Attachment in the Removal of Faecal Bacteria from Macrophyte and Algal
Waste Stabilization Ponds _____ 113

Chapter Nine _____ 131

Effect of Protozoa on Faecal Bacteria Removal in Macrophyte and Algal Waste
Stabilization Ponds _____ 131

Summary in English _____ 141

Summary in Dutch _____ 144

Curriculum vitae _____ 147

ABSTRACT

Waste stabilization ponds are recognized as the solution to domestic wastewater treatment in developing countries. The use of such natural systems is considered to be very important. This is because it is cheap, easy to construct and they do not require high skilled labour. In the developing countries the objectives for wastewater treatment should put emphasis on pathogen removal since most diseases and deaths in these areas are caused by poor sanitation. The efficiency in the removal of pathogens in algal waste stabilization ponds has been found to be very good. The global awareness on resource depletion calls for the use of macrophytes to recover nutrients from wastewater and also to act as an incentive to wastewater treatment. However, the pathogen removal efficiencies of macrophyte-based stabilization ponds are not well known. An understanding of the mechanism involved could be used to improve on the technology and maximize the benefits through effective operation and maintenance practices.

The determination of how these macrophyte ponds will function in a tropical developing country like Ghana including the environmental conditions within the ponds and the mechanisms associated with pathogen removal and die-off were undertaken through this PhD research. Experiments were conducted on batch-scale, bench-scale and pilot-scale continuous flow systems in Ghana and Colombia using water lettuce (*Pistia stratiotes)* and duckweeds (*Lemna paucicostata* and *Spirodela polyrhiza*). The main experimental set up consisted of 3 parallel pond systems operating in series of four ponds comprising of water lettuce, duckweed and algae. This was located in at the Kwame Nkrumah University of Science and Technology, Kumasi, Ghana.

Environmental condition measurements showed that neutral conditions exist in the water lettuce and duckweed ponds except when the pond systems were operated using low strength sewage, a low pH of about 4 was observed in the water lettuce pond system. Low dissolved oxygen levels were associated with the macrophyte ponds. In the algal ponds, high pH of >10 and DO concentrations above saturated levels prevailed.

Performance on continuous flow systems showed that organic matter and nitrogen compounds were effectively removed in the macrophyte ponds. The algal ponds also performed well but the macrophytes were more efficient. The removal of phosphorous was rather poor in all 3-pond systems. Performance studies conducted using low sewage strength showed effective removal of faecal coliforms in the algal ponds. Poor removal of faecal coliforms was observed in the macrophyte ponds. It was however observed, that the middle portions of the last pond in the water lettuce ponds had low numbers of faecal coliforms equal to values obtained in the final effluent from the algal pond system. More than 99% of the faecal coliforms were also found in the sediments of all ponds. There was however no difference among the 3 pond systems for the removal of enterococci. In a medium sewage strength performance studies, removal of *E. coli, Salmonella* and enterococci were the same in all 3-pond systems with no significant ($p<0.5$) differences during a one-year monitoring phase. Water lettuce pond system had poor removal of coliforms and other enterobacteria. All the faecal bacteria removals in the duckweed pond system were comparable to that of the algal pond system during this period. During the intensive measurements study however, (i.e. when the macrophytes were most healthy and the algal ponds were bright green), both macrophytes pond systems performed better in the removal of coliforms and other enterobacteria than the algal ponds.

Total biomass yield in the water lettuce and duckweed pond systems were respectively 503.6 and 286.6 tons/ha/year. Protein content in the macrophytes was highest in the second pond with water lettuce leaves having 22% and duckweed fronds 34%. Duckweed fronds effectively covered all pond surface and completely eliminated mosquito breeding in the ponds.

In pH effects experiments, faecal coliforms were found to be susceptible to high alkaline pH while enterococci were found to be susceptible to low of pH <5. Sunlight was found to inactivate faecal bacteria and promote die-off.

In pH fluctuation experiments, fluctuations in pH had higher die-off rates than stable pH incubations for *Salmonella* and other enterobacteria. Stable alkaline pHs were more detrimental than fluctuating pHs from alkaline to neutral to *E. coli* and coliforms in domestic wastewater and *E. coli* except fluctuating pH of 4-9 and 7-9, which had higher die-off rates than stable pHs of 4, and 9 for *E. coli*. At pH 5, low die-off rates were recorded for *E. coli* and *Salmonella*. For enterococci, fluctuations in pH were more detrimental than stable pH incubations. Extreme pH treatment of 4, 10, and 11 were found to be most detrimental to all the faecal bacteria used in this study. This study showed that pH fluctuations, extremes of pH and presence of other microbes might all contribute to faecal bacteria die-off in domestic wastewater treatment plants depending on bacteria species.

Attachment of bacteria occurs readily on most available surfaces. The question addressed in this study is whether this mechanism plays a role in pathogen removal in macrophyte and algal waste stabilization ponds. An attempt was made to answer this question in trials on a batch-scale, a bench-scale continuous flow system in Ghana and by using a pilot-scale continuous flow system in Colombia, South America. The results showed that faecal bacteria attach to walls of containers holding wastewater, water lettuce roots and leaves, duckweed fronds and algae. When the die-off rates and mass balance of faecal bacteria on various surfaces in batch-scale incubations were studied, die-off was observed immediately after attachment. Higher die-off was observed in surfaces in the algal ponds. Most of the viable bacteria were found attached to water lettuce roots and to suspended algae (over 70%). Harvesting of macrophytes removed <1% of viable faecal bacteria in continuous flow ponds in Ghana and in Colombia. In comparison to percentage of faecal bacteria attachment to surfaces with total viable bacteria numbers, attachment was substantially found to contribute to faecal bacteria removal. Attachment and subsequent settling of suspended solids contribute to prolonged retention of faecal bacteria in stabilization ponds, and as such provides the conditions for die-off.

The effect of the presence of protozoa on faecal bacteria removal and protozoa population densities were assessed. Protozoa were found in both macrophyte and algal pond systems and mostly in sediments and surface at times. Protozoa found were mainly flagellates, ciliates and a few amoebae. The algal ponds had the highest number of protozoa, followed by water lettuce and duckweed ponds respectively. The algal ponds were dominated by *Euglena*. Other algae such as *Chlorella, Chlorococcum, Phacus, Ulothrix* and some diatoms were also found but in small quantities. In the presence of protozoa, removal of *E. coli*, coliforms and *Salmonella* bacteria were ($p<05$) significantly faster than in the absence of protozoa in the water lettuce ponds. In the duckweed and algal ponds, for all the faecal bacteria studied, there was no significant difference in the removal of faecal bacteria in the presence and absence of protozoa.

Chapter One

Introduction

Introduction

Human excreta may contain pathogens (Feacham *et al.*, 1983) and therefore must be properly managed in any community. Lack of adequate treatment and disposal will lead to a high morbidity and mortality. According to WHO (2005), about five million people die annually due to lack of adequate sanitation and about 3 billion people lack access to adequate sanitation. Most of these people live in the developing world.

The millennium development goals (MDGs) seek to decrease the number of people without access to adequate sanitation by 50% by the year 2015 (Cosgrove and Rijsberman, 2000). In order to meet these goals about 460,000 people must be provided with improved sanitation daily (WHO, 2005). To achieve worldwide coverage of water supply and sanitation in 2025, a worldwide investment of $200 billion per year (assuming that conventional technologies are used) is required. These cost estimates do not include full coverage of wastewater management (estimated at $70-90/cap/y), which would raise the total worldwide annual investment requirement to some $600-800 billion (Bos *et al.*, 2005).

The huge sums of money required to improve sanitation seems unrealistic for many developing countries to meet. It is necessary therefore to consider low-cost options, which will be able to meet treatment objectives of reducing public health risks (pathogens), organic and nutrient concentrations. These treatment objectives are necessary to ensure that the natural receiving environments will be able to assimilate these contaminants without undue stress. The treated wastewater could also be used in agriculture. In many urban centres world wide, drinking water is used to transport human excreta to central points for treatment. Even though the corresponding infrastructure of flush toilets, sewer systems, centralized hard ware and energy demanding wastewater treatment plants, add to the cost of treatment, many developing countries have adopted the same concept without examining the consequences of such technologies.

In developed countries water from toilets, bathing, washing, (pre-treated) industrial effluents, and storm water are usually collected in a single pipeline and channelled to a centralized treatment plant. This wastewater is cumulatively known as sewage. The wastewater from the toilets is known as black water while that from kitchen and bathrooms is known as grey water. The two types of wastewater are directed in a wastewater channel and together referred to as domestic wastewater.

Several technologies are used to treat domestic wastewater. These can be classified into two groups: conventional and non-conventional treatment plants. The former has high-energy requirements. The later is solely dependent on natural purification processes. The conventional systems of wastewater treatment include trickling filters, activated sludge systems, biodisc rotators and aerated lagoons.

The non-conventional systems, which are also called eco-technologies (Nhapi and Gijzen, 2005) include constructed wetlands and waste stabilization ponds. Among these technologies, the widely recommended ones for developing countries are the waste stabilization ponds (WSPs).

The general aim of wastewater management is to collect wastewater from the homes, treat and discharge into the environment without causing undue stress to the environment. In Ghana, most of the wastewater is stored and partially treated in septic tanks. Only two towns Akosombo and

Tema municipalities are completely sewered. Partial sewerage networks are however found in Accra and Kumasi (Figure 1).

Figure 1 Map of Ghana

The septage or sludge accumulation from the septic tanks is usually discharged into the environment untreated. Grey water is combined with storm water and discharged into natural drains without prior treatment. The water from the drains and raw blackwater is used for agricultural purposes (Figure 2).

The objective of wastewater treatment is to reduce organic loads, nutrient levels and to eliminate pathogens to render the effluent safe to handle by the receiving environment. The water is treated to avoid undue stress to the environment backed by regulations for compliance. With increased awareness of health and environmental impacts associated with wastewater mismanagement, more stringent guidelines have been produced making the cost of treatment even more expensive. One of the major concerns is the presence of pathogens in wastewater. In the conventional treatment systems, pathogens are not efficiently removed. As a result, chemical disinfectants are widely used to eliminate the pathogens in the treated effluents.

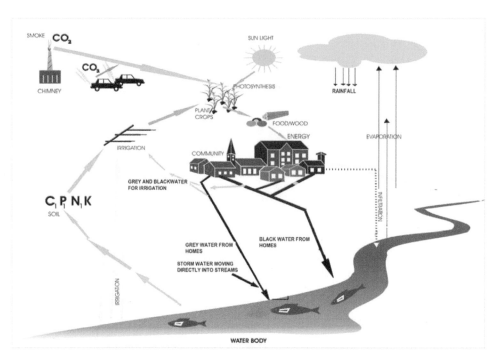

Figure 2 Current practice of domestic wastewater management in Ghana

The current practice of wastewater management is giving way to a new concept. In the current concept, the effluent from the wastewater treatment plants goes into the receiving water body without reuse and recycling. There is also no rational use and resource recovery (Figure 2). In the developing countries, most wastewater treatment systems have broken down because of lack of incentives. If macrophytes are used in the treatment of wastewater and harvested for economical activities they could become a major source of income and help balance the human ecosystem in many urban centres at a fast rate (Gijzen 2001). By using ecotechnologies, all waste generated in a community could be treated and this goes back into the soil or is used in irrigation for crop production or for other purposes such as woodlots. Energy is generated during the process of anaerobic digestion. This could then go back into the community for use. Effluent from the wastewater treatment plants goes back to the homes to be used for flushing toilets. In this way drinking water, which is very expensive to treat is not used for flushing toilets. This ensures that portable water is rationally used (Figure 3). This is the new concept of wastewater management. Alternatively waterless toilets could also be used.

The new way of managing wastewater should not only look at meeting guidelines but also should consider rational use of resource, resource recovery, reuse and recycle of matter as practiced in natural ecosystems in view of rapid population growth and resource depletion (Figure 3). This is a sustainable way for wastewater management (Bos *et al.*, 2005).

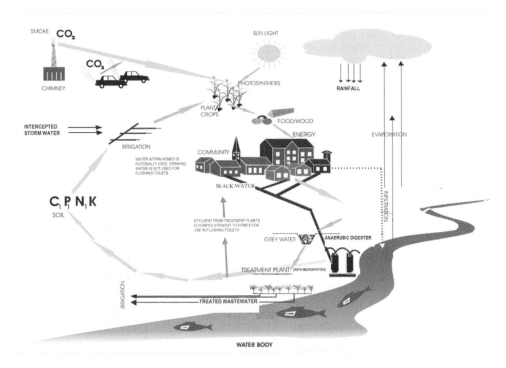

Figure 3 A new concept of wastewater management (modified from Otterpohl *et al*, 1998).

Wastewater treatment and cost comparisons

Waste stabilization ponds (WSP) are defined in this paper as wastewater containing ponds relying on natural processes based on algae and bacteria for organic, nutrient and pathogen loads removal. WSPs are without doubt the most important and effective method of sewage treatment in developing countries. They are the least expensive wastewater treatment system (Table 1) and are considerably more efficient in destroying pathogenic bacteria than other treatment plants such as trickling filters and activated sludge systems (Table 2).

When the waste stabilization pond is covered with floating macrophytes it becomes a macrophyte pond. When it lacks the macrophytes it is known as algal pond (Zimmo, 2003). Stabilization ponds have generally been recognised as low-cost and simple technologies for the effective treatment of wastewater in tropical developing counties. Apart from algal ponds, macrophyte ponds have recently been proposed (Polprasert, 1996; Zimmo 2003; Gijzen, 2001). The reason for the introduction of macrophytes in pond systems is to provide incentives for income generation through resource recovery and re-use of treated effluent for agricultural or aquaculture purposes.

Algal ponds are waste stabilization ponds that are naturally infested with algae of which several species may be involved. Some of the species are *Nostoc*, *Euglena* and *Scenedesmus*. Algal ponds are easy to construct and manage. They also achieve satisfactory levels of treatment (Lansdell, 1987; Wang, 1991). They are efficient in the removal of pathogens Table 2 (Feacham *et al.*, 1983) although the efficiency is not always 100%. Algal ponds, however, have certain shortcomings such as the large land area requirements and the high concentrations of suspended algal cells in the effluent, which can make re-use of the wastewater problematic. Use of effluent from algal ponds can result in clogging of drip irrigation pipes. Harvesting of algae may require complex technologies often incompatible with the technological systems in many developing countries (Pearson *et al.*, 1996). Besides, the N and P removal mechanisms from conventional stabilization ponds are not well understood.

Table 1 Estimated total annual and unit costs for alternative treatment processes with a design flow of 4.5million L/day

Process	Initial Capital Cost (US$)[a,b]	Annual Cost (US$)			Unit Cost (cent/ 1000L)[b]
		Capital[c]	O&M[d]	Total	
Imhoff tank	380,000	41,720	15,500	57,270	3.5
Rotating biological disk	800,000	87,832	57,680	145,512	8.9
Tricking filter	900,000	98,811	58,480	157,291	5.6
Activated sludge with					
External digestion	1,000,000	109,790	74,410	184,200	11.2
Internal digestion [e]	500,000	54,895	48,800	103,695	6.3
Stabilization pond [f]	250,000	27,447	23,680	51,127	3.1
Land disposal [g]					
basic system	340,000	37,328	41,540	78,869	4.8
with primary treatment	940,000	103,302	81,540	184,742	11.2
with secondary treatment	1,240,000	136,139	115,950	252,089	15.5
Land disposal [h]					
basic system	200,000	21,958	25,100	47,058	2.9
with primary treatment	800,000	87,832	5,100	152,932	9.3
with secondary treatment	1,000,000	109,790	99,510	209,300	12.7

a) *estimated average cost.* b) *based on engineering news-record building cost index of 1900.*
c) *capital recovery factor = 0.10979 (15 years at 7%.)*
d) *average values for variations in processes* e) *external aeration, aerated lagoon, oxidation ditches.*
f) *high-rate aerobic, facultative, and anaerobic* g) *irrigated and overland flow.* h) *infiltration-percolation.*
Adapted from Salvato 1992

In algal waste stabilization ponds, drastic changes have been observed in environmental conditions in the ponds like strong diurnal pH fluctuations, high pH peaks, and alternating high and low dissolved oxygen concentrations. These conditions may be detrimental to pathogens (Moeller and Calkins, 1980; Pearson *et al.*, 1987 a,b). Solar energy and algal photosynthesis have been identified as being responsible for generating these conditions (Parhad and Rao, 1972; 1974; Curtis, 1990; Curtis *et al.*, 1992a and b; Moeller and Calkins, 1980). The adverse conditions of fluctuating pH and high O_2 and pH levels, characteristic of algal ponds, are not likely to occur in macrophytes systems since in the latter there is no direct penetration of solar radiation into the water phase.

Reliance on floating macrophytes (aquatic weeds) to treat wastewater in stabilization ponds is gaining recognition worldwide because of the ability of these macrophytes to remove nutrients from wastewater and the potential use of the aquatic plants for agricultural purposes and other economic activities (Von Sperling, 1996; Zimmo, 2003). Macrophyte ponds (MP), however, do not readily permit direct sunlight penetration and may also serve as breeding grounds for vectors such as mosquitoes. The ability of this treatment system to remove pathogens is very crucial especially if the macrophytes are going to be used for fish farming activities (Gijzen and Khonker, 1997) or for animal feed (Nhapi and Gijzen, 2005).

Table 2 Summary of pathogen removal efficiencies of various sewage treatment processes

Organisms	Parameters	Primary sedimentation	Trickling filter with, sludge digestion and sludge drying	Activated sludge with primary and secondary sedimentation, digestion and sludge drying	Oxidation ditch with sedimentation and sludge drying	Waste stabilization ponds with minimum RT=25days	Septic tanks
Enteric viruses	%Removal	0 – 30	90 – 95	90 – 99	90 – 99	99.99 – 100	50
	Final sludge	Contaminated	Contaminated	Contaminated	Contaminated	-	Contaminated
Salmonella	%Removal	50 – 90	90 – 95	90 – 99	90 – 99	99.99 – 100	50 – 90
	Final sludge	Contaminated	Contaminated	Contaminated	Contaminated	-	Contaminated
Shigella	%Removal	50 – 90	90 – 95	90 – 95	90 – 99	99.99 – 100	50 – 90
	Final sludge	Contaminated	Contaminated	Contaminated	Contaminated	-	Contaminated
Escherichia coli	%Removal	50 – 90	90 – 95	90 – 95	90 – 99	100	50 – 90
	Final sludge	Contaminated	Contaminated	Contaminated	Contaminated	-	Contaminated
Vibrio cholerae	%Removal	Contaminated	Contaminated	Contaminated	Contaminated	100	Contaminated
	Final sludge	Contaminated	Contaminated	Contaminated	Contaminated	-	Contaminated
Leptospira	%Removal	10	10	10	10	100	100
	Final sludge	safe	safe	safe	safe	-	safe
Entamoeba histolytica cysts	%Removal	10-50	50	50	50	100	0
	Final sludge	Contaminated	safe	safe	safe	100	Contaminated
Hookworm ova	%Removal	50	50-90	50-90	50-90	-	50 – 90
	Final sludge	Contaminated	Contaminated	Contaminated	Contaminated	100	Contaminated
Ascaris ova	%Removal	30-80	70-100%	70-100	70-100	100	50 – 90
	Final sludge	Contaminated	Contaminated	Contaminated	Contaminated	-	Contaminated
Schistosoma ova	%Removal	80	50-90	50-99	50-99	100	50 – 90
	Final sludge	Contaminated	safe	safe	safe	-	Contaminated
Taenia ova	%Removal	50-90	50-90	50-90	50	100	50 – 90
	Final sludge	Contaminated	Contaminated	Contaminated	Contaminated	-	Contaminated

Adapted from Feacham *et al.*, 1983

There are several species of floating aquatic weeds that can be used in MP systems of wastewater treatment. Their selection should, however, be based on the local conditions in which they will be utilised. In Ghana, duckweed (*Lemna* sp. and *Spirodela* sp.) and water lettuce (*Pistia stratiotes*) appear to be good plants for use in such ponds because they are readily available and pose no serious threat to the local aquatic environment. The introduction of macrophytes into stabilization ponds seeks to address the problem of resource recovery by converting nutrients available in wastewater into reusable forms as incentive to wastewater treatment but not primarily on pathogen removal.

Pathogen removal is the most important objective in wastewater treatment in developing countries, because of the potential hazards caused by enteric diseases in these countries. The use of macrophyte-based ponds for resource recovery calls for a critical examination of the mechanisms involved in pathogen removal. Knowledge on pathogen removal mechanisms could provide the basis for improvement in the design and operation and maintenance of waste stabilization ponds.

Conflicting results exist in the efficiency of MP and AP systems regarding pathogen removal from wastewater as elaborated below. Studies with indicator organisms indicate that the removal of such organisms is more efficient with AP systems than with MP systems (van der Steen *et al*., 2000). Other reports indicate that MP systems are more efficient than AP systems (Mandi *et al*, 1993). These studies were conducted in different climatic regions and the wastewater characteristics were different. Besides, judgement on the efficiency of pathogen removal based only on the indicator organism *Escherichia coli* may not be appropriate since other pathogenic organisms exist in wastewater and these also must be monitored in comparative studies before coming to firm conclusions.

Conventional systems for wastewater treatment depend on the use of chlorine to achieve complete destruction of pathogens. However, the chlorine must be removed with sulphide if the effluent is to be discharged into water bodies containing fish. Since chemical use in the treatment of wastewater creates environmental problems and is also expensive, natural alternatives to chemical use such as AP and MP systems should be considered. The pathogen removal mechanisms in these natural wastewater treatment systems, however, are poorly understood, and need to be studied.

The objective of this PhD study is to conduct a comparative study on pathogen removal mechanisms in AP and MP systems under tropical conditions. Knowledge of the mechanisms involved in pathogen removal could serve as the basis for improving the design, construction and operation of these promising wastewater treatment systems.

Characteristics of wastewater

A typical domestic wastewater contains several substances (Metcalf and Eddy, 2003). Wastewater could be classified into 3 categories as weak, medium and strong wastewater as shown in Table 3.

The characteristics of wastewater depend on many variables, including the country, the type of diet, health status and water use patterns (Table 4). The strength of sewage depends mostly on the amount of water used per capita per day. In countries where water is scarce the wastewater tends to be strong.

Table 3 Typical composition of domestic wastewater

Parameter	Weak	Medium	Strong
Total solids, (TS)	350	720	1200
Dissolved solids, total (TDS)	250	500	850
Fixed	145	300	525
Volatile	105	200	325
Suspended solids (SS)	100	220	350
Fixed	20	55	75
Volatile	80	165	275
Settlable solids	5	10	20
BOD_5, 20^0C	110	220	400
Total organic carbon (TOC)	80	160	290
COD	250	500	1000
Total Nitrogen	20	40	85
Organic Nitrogen	8	15	35
NH_4^+	12	25	50
NO_3^-	0	0	0
NO_2^-	0	0	0
Total phosphorus	4	8	15
Organic phosphorus	1	3	5
Inorganic phosphorus	3	5	10
Chlorides	30	50	100
Sulphate	20	30	50
Alkalinity(as $CaCO_3$)	50	100	200
Grease	50	100	150
Total coliforms	10^6-10^7	10^7-10^8	10^7-10^9
Volatile organic compounds	<100	100-400	>400

Metcalf and Eddy (2003). *All concentrations are in mg/L with the exception of total coliforms No./100mL and volatile organic compounds, $\mu g/L$.*

Table 4 Typical composition of domestic sewage in some countries

Determinant	Austria	Netherlands	Morocco	Turkey
COD (mg /L)	526	450	928	656
BOD (mg /L)	285	171	353	212
SS	nd	237	397	150
N	44	42	nd	40
P	7.1	6.7	nd	nd

Adapted from Pons et al., 2004, nd = not determined

For comparison, the characteristics of wastewater at the Kwame Nkrumah University of Science and Technology and Asafo, a suburb in Kumasi, Ghana are shown in Table 5. The wastewater at Asafo is unusually strong because the inhabitants allow several people to visit the toilet before a single flush in order to cut down on cost (Kuffour, personal communication). The wastewater from the University is used throughout the research described in this thesis.

The characteristics of wastewater, with particular reference to pathogen counts, also depend on the health status of the people. Human beings, infected with a particular disease may discharge pathogenic organisms into wastewater and affect the composition of the wastewater.

Table 5 Characteristics of wastewater generated at KNUST and in Asafo.

Parameter	KNUST, grey water	KNUST, sewage	Asafo, sewage
pH	7.5±0.2	8.2 ± 0.1	8.0 ±0.0
Temperature (°C)	29.2±0.7	28.5 ± 0.2	27.7 ±0.5
Dissolved oxygen (mg/L)	2.7 ± 0.8	0.6 ± 0.1	1.0 ± 0.5
TSS (mg/L)	212 ± 21	595 ± 221	1152 ± 345
BOD (mg/L)	198 ± 33	310 ± 76	1007 ± 378
COD (mg/L)	399 ± 108	667 ± 103	2540 ± 641
Ammonia-N (mg/L)	8.4 ± 1.8	120.8 ± 0.5	148 ± 9.5
Nitrate-N (mg/L)	0.7 ± 0.1	1.7 ± 0.6	18.7 ± 9.0
Nitrite-N (mg/L)	0.00 ± 0.00	0.02 ± 0.04	0.16 ± 0.11
Phosphorus (mg/L)	11.8 ± 4.0	11.2 ± 1.9	17.0 ± 9.9
Total coliforms/100mL	$2.7 \times 10^7 \pm 0.6 \times 10^8$	$5.4 \times 10^8 \pm 7.1 \times 10^9$	$4.2 \times 10^{10} \pm 5.9 \times 10^{10}$
Faecal coliforms/100mL	$1.5 \times 10^6 \pm 0.2 \times 10^6$	$3.8 \times 10^7 \pm 3.5 \times 10^7$	$5.9 \times 10^8 \pm 5.4 \times 10^8$
Helminth eggs (No/L)			
Ascaris sp.	2333 ± 1527	7000 ± 4,044	62,116 ± 45000
Tapeworms	6500 ± 658	14,200 ± 608	25,306 ± 6985
Fasciola sp.	1300 ± 282	2400 ± 670	3473 ± 225
Trichuris sp.	Nil	Nil	1236 ± 108
Schistosoma sp.	Nil	Nil	373 ± 71
Metals (mg/L)			
Fe	0.37±0.08	0.97±0.01	0.01 ± 0.04
Hg	0.4±0.03	0.2±0.13	<0.01
Ca	2.81 ± 0.01	2.44 ± 0.01	2.72 ± 0.02
Cd	0.01 ± 0.00	<0.01	0.01 ± 0.00
Pb	<0.01	<0.01	<0.01
Cu	<0.01	0.03 ±0.03	0.01 ± 0.03
Zn	0.03 ± 0.00	0.02 ±0 .00	0.1 ± 0.00
Al	<0.01	0.70 ± 1.67	0.6 ±0.80
Mn	0.04 ± 0.01	0.13 ± 0.02	<0.01
Mg	6.1 ± 0.40	4.8 ± 0.10	4.5 ± 0.10

Adapted from Awuah *et al.*, 2002, ± *Standard deviation.*

The pathogen levels in the Asafo wastewater are higher than the levels in the KNUST sewage. Besides, the number of pathogens in the wastewater in both communities (Asafo and KNUST) are quite high, partly because of the strength (many people use the toilets before a single flush) and most probably their low educational background resulting in poor hygienic practices. This may not be different from other parts of the country and necessitates treatment of all domestic wastewater produced in Ghana to improve on existing sanitary conditions.

Pathogenic organisms found in domestic wastewater

Pathogenic organisms in wastewater are categorised into viruses, bacteria, protozoa, fungi and helminths. Some of these organisms found in wastewater can cause diseases of the gastrointestinal tract such as typhoid and paratyphoid fever, dysentery, diarrhoea and cholera.

Bacterial pathogens are very harmful and are the cause of many deaths in areas with poor sanitation, especially in the tropics. Although bacterial pathogenic organisms are the most numerous, they are in no means the only pathogens in wastewater. Table 6 gives persistent pathogenic microorganisms found in wastewater.

Table 6 Pathogenic organisms persistent in domestic wastewater

Organism	Disease
Helminths	
Ascaris lumbricoides	Ascariasis (roundworm infestation)
Enteribium vermicularis	Enterobiasis, (pinworm)
Fasciola hepatica	Fascioliasis (sheep liver fluke)
Hymenolopis nana	Hymenolopiasis (dwarf tapeworm
Taenia saginata	Taeniasis (beef tape worm)
Taenia solium	Taeniasis (pork tapeworm
Trichuris trichura	Tricuriasis (whipworm)
Protozoa	
Balantidium coli	Balantidiasis (diarrhoea)
Cryptosporidium parvum	Cryptosporiasis (diarrhoea)
Entamoeba histolytica	Amoebic dysentery
Giardia lamblia	Giardiasis (diarrhoea, nausea, indigestion)
Bacteria	
Escherichia coli	Gastroenteritis (diarrhoea)
Legionella pneumophila	Acute respiratory illness
Leptospira sp.	Jaundice, fever.
Salmonella typhi	Typhoid fever
Salmonella (~2100serotypes).	Food poisoning
Shigella sp.	Bacillary dysentery
Vibrio cholerae	Cholera
Yersinia enterolitica	Diarrhoea
Viruses	
Adenovirus (31 types)	Respiratory disease
Enteroviruses (polio, echo and coxsackie)	Gastroenteritis, heart anomalies, meningitis
Hepatitis A	Infectious hepatitis (Jaundice)
Norwalk agent	Gastroenteritis
Parvovirus (3 types)	Gastroenteritis
Rotaviruses	Gastroenteritis

Adapted from Crites and Tchobanoglous (1998)

The ability of a pathogen to cause a disease depends on the dose (Table 7) and the susceptibility of the host. In addition to the organisms provided in Table 6, several other viruses are also found in wastewater including Human Immune Deficiency Syndrome. Fortunately, HIV does not survive in wastewater for a long time (Moore, 1993).

Bacterial species such as *Mycobacterium tuberculosis,* helminths such as hookworm (*Ancylostoma duodenales*) and other intestinal worms like *Schistosoma mansoni, S. japonicum, S. haematobium* and *Daphylobotrium latum* (associated with pets) are persistent in the wastewater where the infection is endemic. Pathogenic fungi found in wastewater are few and they include *Candida albicans* (Feacham *et al.,* 1983; Bitton, 1994).

Experimental results have shown that ten to hundred thousand infectious doses of hepatitis virus are emitted from each gram of faeces of a patient infected with this disease.
About 10^{11} bacterial pathogens are excreted by each person in a day (Bitton, 1994).

Table 7 Concentration and infectious doses of pathogenic organisms, occurring in wastewater

Organisms	Concentration in raw wastewater (No/100mL)	Infectious dose
Helminths		
Ascaris lumbricoides (ova)	$10^1 - 10^3$	1-10
Protozoa		
Cryptosporidium parvum oocysts	$10^1 - 10^4$	1-10
Entamoeba histolytica	$10^1 - 10^3$	10-20
Giardia lamblia	$10^3 - 10^4$	<20
Bacteria		
Total coliforms	$10^7 - 10^9$	(unknown)
Faecal coliforms	$10^6 - 10^8$	$10^6 - 10^{10}$
Clostridium perfringens	$10^3 - 10^5$	$1 - 10^{10}$
Faecal streptococci	$10^4 - 10^5$	-
Pseudomonas aeruginosa	$10^3 - 10^4$	-
Shigella	$10^0 - 10^3$	10-100
Salmonella	$10^2 - 10^4$	$10^4 - 10^7$
Viruses		
Enteric viruses	$10^3 - 10^4$	1-10 plague forming units

Adapted from Bitton (1994); Crites and Tchobanoglous (1998)

Indicator organisms

The identification of faecal pathogenic organisms in water and wastewater can be extremely time-consuming, expensive and difficult. The coliform group of organisms is therefore used as indicator of bacterial pathogenic organisms (Greenberg *et al.*, 2003). The coliform bacteria include the genera *Escherichia* and *Aerobacter*. The use of coliforms as indicator organisms is complicated by the fact that *Aerobacter* and certain *E. coli* strains can grow in soil (Byappanahalli and Fujioka. 2004). Thus, the presence of coliforms does not always mean contamination with human wastes. Apparently, *E. coli* is entirely of faecal origin and hence widely used as an indicator of faecal contamination organism. However, there are problems associated with the use of *E. coli* as the sole indicator organism. Lesgne *et al.*, (1991) and Mezrioui and Oudra, (1998) found out that *Vibrio cholerae* and *E. coli* behave differently under the same conditions. They found that whereas *V. cholerae* populations increased with pH and temperature, *E. coli* populations decreased and they concluded that a great deal of caution was needed to be used when assessing the health risks of wastewater based on faecal pollution indicators alone. Nascimento *et al.*, (1991) found removal kinetics to be different for different indicator organisms and recommended that multiple indicators should be used in the kinetics of pathogen removal. Other authors also doubt the reliability of *E. coli* (Burkhardt III *et al.*, 2000; Len *et al.*, 2000). This is because the behaviour of the real pathogens may be different from the indicator organisms. Most research work on pathogen removal has been done using *E. coli* and *Enterococcus*.

All these bacteria belong to the family Enterobacteriaceae and occur mostly in the intestines of animals including man. These Gram-negative rods are usually associated with intestinal infections. They are the causative agents of such diseases as meningitis, bacillary dysentery, typhoid, and food poisoning (Table 6).

Techniques for indicator organism isolations

Techniques usually used to evaluate the microbiological safety of water are based on determining its content of "indicator organisms". These are organisms that occur extensively in human wastes, whether or not the person suffers from a waterborne disease. Accordingly, they do not necessarily relate directly to the populations of pathogens but provide an estimate of the

extent to which human wastes have contaminated the water recently. The principal indicator organism used in evaluating water quality is the coliform group of bacteria.
Desirable characteristics of indicator organisms are:

- They should be harmless to humans.
- They should be present in polluted waters when pathogens are, or might be, present.
- The number of indicator organisms in polluted water should be correlated with the probability that pathogens are present.
- They should be present in polluted waters in numbers higher than those of pathogens.
- They should be easy and quick to identify and to enumerate through relatively simple laboratory tests.
- They should not multiply under conditions where pathogens do not multiply.
- They should survive unfavourable environmental conditions longer than pathogens do. (This insures that waters that have been treated to produce low or zero populations of indicator organisms are safe).

Several methods are employed in the determination of indicators organism. The coliform test is used as an indicator of sanitary quality. Faecal coliforms are standard indicators of choice for shellfish and shellfish harvested waters, wastewater treatment plants effluent quality and general pollution trends in surface and ground water. *E. coli* is used to indicate recent faecal contamination. Techniques for the enumeration of coliforms, faecal and *E. coli* include the most probable number (MPN) technique that is a standard statistical multi-step method consisting of presumptive, confirmatory and completed phases using serial dilution in multiple tubes. The use of membrane filters is equally used but when the water or wastewater has a lot of suspended solids it may interfere with the results. Serial dilutions of the sample are made and plated in an absorbent pad embedded in a broth or a solid agar. Continuous developments in faecal bacteria enumerations have resulted in the production of chromocult agar that can isolate pathogens like *Salmonella* simultaneously with *E. coli*, coliforms and other enterobacteria, which cannot be chromogenically differentiated by the medium. The medium has been found to be suitable for faecal bacteria determination in wastewater treatment plants (Byamukama *et al.*, 2000).

Identification of microbial pathogens has also been done with the help of genetic probes (Actis *et al.*, 2003). In this PhD research, MPN, membrane filtration and Chromocult agar isolation methods for faecal bacteria identification were used.

The coliform group of bacteria

Large populations of certain bacteria *(Escherichia coli)* grow in the intestinal tracts of humans and are excreted in faecal wastes. The "coliform group" of bacteria includes *Escherichia coli (E. coli)* and some different types that originate from other sources as well as in human faecal discharges.
The coliform group is used to evaluate the microbiological quality of waters; the number present is interpreted as an indicator of the extent to which that water has been contaminated recently by human faecal discharges. Actually, the presence of coliforms does not establish that pathogens necessarily are there. Some coliforms are capable of causing disease, but most are not viewed as pathogenic; however, the number present does provide a measure of the probability that waterborne pathogens might be there.

Although the coliform group meets some of the criteria just listed, it does have certain limitations. For example, many bacteria in the group do not originate from the intestinal tracts of humans and, therefore, really bear little or no relationship to the opportunity for pathogens to be present. Also, coliforms often can multiply in treatment facilities or watercourses (Gibbs *et al.*, 1997). This can produce high populations without corresponding sanitary significance, because pathogens are far less likely to multiply under those same environmental conditions.

Coliforms usually outlast most of the common pathogenic bacteria during natural die-off in streams or in treatment facilities, but they do not always survive viruses and several other organisms. There is concern that the reduction of coliforms to low levels, or even their elimination, may not always be accurate in indicating that the water is safe. For example, *Giardia lamblia* often is present in unacceptable numbers in drinking waters, even with negative or low coliform tests (http://water.sesep.drexel.edu/outbreaks/Sydney_5/r5vol1.htm).

In spite of the shortcomings, the coliform test has served well to evaluate the risks of transmitting waterborne diseases. It has been used for over 70 years as a key part of many successful programs around the world for controlling the spread of waterborne diseases. Although its accuracy and precision leave something to be desired, it still has provided a valuable and generally reliable technique for evaluating the safety of drinking water and warning of unsafe microbiological conditions when they occur. No other analytical method has been devised yet that does the job better.

Even though these indicator organisms are supposed to be non-pathogenic, *E. coli* and enterococci can become pathogens and have caused outbreaks of enteric diseases (Health News, 2004). Some are also antibiotic resistant and this is considered as a threat to public health (Cetinkaya *et al.*, 2000).

Review on pathogen removal mechanisms

Microbial cells are in a dynamic state, adapting readily to shifts in environmental conditions. Examples are modification of enzyme synthesis to take up growth limiting nutrients, modulation of uptake rates for nutrient available in excess, re-routing of metabolic pathways to avoid possible blockages due to specific nutrient limitation and co-ordination of synthetic rates to maintain balanced growth (Rozac and Colwell, 1987). These adaptive processes vary from organism to organism and can prevent the removal of pathogens under different conditions. Both pathogens and non-pathogenic microbes in the external environment other than the host are subjected to several processes and factors, which bring about their death or removal. Microorganisms are removed in nature through sedimentation, attachment, predation, filtration, allelopathy, direct sunlight and harsh environmental conditions such as extreme temperatures, pH, DO and dissolved solids. Mechanisms affecting pathogen removal may be grouped in three broad categories. These are biological and physical processes and chemical conditions all operating to bring about microbial removal and die-off.

Removal in this context is the removal of pathogens from a treatment system through mechanical or engineering designs such as sedimentation through tanks, direct kill due to harsh environmental conditions created in waste treatment system, or predation of the pathogen by other organisms.

Physical factors and processes

Basically, the physical processes that affect pathogen removal in wastewater are filtration and sedimentation. Attachment has been added in this section because it is closely linked to sedimentation. Factors such as temperature and sunlight have been included in this section.

Temperature

Generally, optimal growth of bacteria occurs within a fairly narrow range of temperature, although the organism may be able to survive within much broader limits. Temperatures above the optimum typically have a more significant effect on growth rate than temperature below the optimum; it is reported that growth rate of bacteria doubles with approximately every 10^0C rise in temperature (expressed as $Q_{10}=2$) until the optimum is reached (Metcalf and Eddy, 2003). More research work has considered faecal bacteria removal as a much more complex mechanism involving interactions between the physical, chemical and biological systems present in the lagoon although temperature clearly remains an important parameter. For example, Polprasert et al., (1983), Pearson et al., (1987a, b), Barzily and Kott (1991), and Mezrioui et al., (1995) have all observed that removal of faecal coliforms is increased with increase in temperature. However, Mara and Pearson (1986) pointed out that the relationship between die-off and increasing temperature must be indirect, since high levels of removal were found in tertiary lagoons in comparison to anaerobic and facultative lagoons operating at the same temperature. The indirect factors could be pH shifts, oxygen (radicals etc.), which might have played a role.

Other scientists (WPCF, 1990; USEPA, 1988) showed that the minimum water temperature needed to enable the use of stabilization ponds in wastewater treatment is 7^0C. At such temperatures pathogen removal will be achieved as conditions, which are created, become detrimental to pathogens.

Sunlight

Sunlight has a lethal effect on coliforms and the rate of die-off is proportional to the intensity of sunlight (Polprasert et al., 1983; Gersberg et al., 1987; Curtis et al., 1992a; 1994). Sunlight effects on faecal bacteria removal have mostly been done in stabilization ponds.

The reason for the detrimental effect of faecal coliforms in the stabilization ponds is the ultra-violet (UV) radiation (Moeller and Calkins, 1980; Whitlam and Codd, 1986; Curtis et al., 1994). Meckes (1982) reported that coliform isolates, resistant to streptomycin, tetracycline and chloramphenicol, were killed or inactivated by UV light. To be precise, Moeller and Calkins (1980) and Davies-Colley et al., (1997) found the range of UV light with wavelength range of 290-320nm, as very important in the destruction of coliform bacteria. They found that UV-B light reduces faecal coliforms by 99% in lagoons.

The fact that sunlight affects pathogen removal calls for the elimination of attenuation effects in the ponds. When an influent is discharged at a great depth, the die-off of bacteria can slow down considerably because sunlight cannot penetrate deeply enough.

Mara and Pearson (1986) also stated that algae are found in all tertiary lagoons. Since algae are photosynthetic, they contain large amounts of pigments. According to Gray (2004), the amount of solids synthesised, largely algal biomass, is proportional to the degree of illumination. However, the more the amounts of algae present, the lower the sunlight penetration. Laboratory studies in experimental ponds have demonstrated that the degree of illumination has no effect on heterotrophic bacteria activity and that BOD removal is related to heterotrophic bacteria density (Gray, 2004). The presence of algae is known to cause increase in pH and dissolved oxygen concentrations.

Davies-Colley et al., (1999) concluded, based on their experiments on light, pH and oxygen relationships using F-DNA, F-RNA viruses, enterococci and E. coli, that there are three light mechanisms operating in stabilization ponds:

- *Mechanism 1* is caused by the direct UV-B light, leading to an oxygen independent damage to F-RNA and F-DNA viruses. Under low light doses repair may occur.
- *Mechanism 2* is caused by photo-oxidative damage by oxygen reactive species produced by substances present in the cells. A range of internal targets may be damaged by single strand DNA breaks. This process is weakened when the light intensity is low.
- *Mechanism 3* is the photo-oxidative damage by exogenous photosensitizers notably dissolved humic substances, which absorb a wide range of wavelength of the sunlight, decreasing in efficiency as wavelength increases.

The type of mechanisms for the removal of the pathogens studied by the authors was dependant on the type of pathogen.

Continuous diurnal measurements of *E. coli* levels in wastewater indicated a decrease during the day (after exposure to sunlight) and an increase after sunset (Almasi and Pescod, 1996). Gibbs *et al.,* (1997) explained the above observation to be due to revival of pathogens, which have been inactivated after exposure to harsh conditions (high pH) when favourable night conditions return.

Liltved and Landfold (2000) support the existence of mechanism 1 and explained further that DNA damage is caused by the absorption of UV-B (280-320nm) by DNA leading to the formation of photoproducts (pyrimidine dimers). These substances further cause cessation of growth and ultimately death. For photo-oxidation damage, in mechanism 2, Liltved and Landford (2000) believed that the indirect effect of photo-oxidation is caused when sunlight is absorbed by a sensitizer, which enters an excited state capable of causing cell damage. In the presence of oxygen, reactive species of singlet oxygen, superoxides, hydrogen peroxide and hydroxyl radicals can be produced (Zepp, 1988; Curtis *et al.,* 1992a). Since the cytoplasmic membrane is a preferable site for the reactive species rendering it susceptible to osmotic attack, they concluded that it is the destruction of the cell membrane that leads to the death of the bacteria under sunlight conditions.

Under dark conditions, *E. coli* is capable of photo repair (Harm, 1968), which can take place in fractions of minutes (Liltved and Landfold, 1996). This means that sampling of effluent from stabilization ponds and other experiments evaluating the performance of bacteria removal should be done after photo repair, usually in the early hours of the morning before full sunlight activity takes place. To summarise, the damage caused by sunlight is due to the toxic oxygen radicals produced, which affect the cell membrane and make it more susceptible to simple factors such as osmotic shock and pH. The complexity of sunlight effects needs further elucidation. The effect of nutrient load and conductivity may all affect sunlight effects and this should be investigated further. Earlier works by Orlob, (1956) and recent works by Van der Steen *et al.,* (2000) and Liltved and Landfold, (2000) suggest that the addition of nutrients like glucose and solutes like NaCl respectively, increase the survival chances of bacteria under both light and dark conditions. This may explain the contradictions common in the literature since other factors affecting bacterial removal are usually not considered.

Filtration
Filtration consists of straining out particles larger than the pore size of the filter. In filtration, there is sometimes adhesion of smaller particles to large ones, which are then filtered. Filtration by the soil medium and attached biofilm has been suggested as one of the mechanisms involved in pathogen removal in wetlands (Vincent, 1994). This is also expected to occur in MP through attachment on submerged plant surfaces.

Attachment/Adhesion

Attachment of pathogens may take place on the surface of plants as well as on the inner walls of the container of the wastewater and to solid matter that sinks to the bottom as sediment. Almost all pathogens have some means of facilitating their attachment to host tissues. According to Tortora *et al.*, (2003) adherence is a necessary step in the pathogenicity of most pathogens. The attachment between pathogen and host takes place by means of surface projections on the pathogen, called ligands (adhesins), and complementary surface receptors on the host cells. The attachment is mostly due to the polymers produced by flagella and fimbriae or pili (Marshal, 1973; Rheinheimer, 1992; Droopo and Jonash, 1980). In some cases, the ligands are associated with other microbial surface structures such as pili. For instance, an *E. coli* strain, which is a commensal in the human intestinal tract, attaches itself to intestinal epithelial cells by means of ligands situated on the pili. *Enterococcus faecalis* adheres to the surface of the tooth enamel by producing an extracellular polysaccharide whereas the eggs of some helminths have hooks which act as adhesive structures (Fletcher, 1996; Tortora *et al.*, 2003).

Attachment may play a role in sedimentation; due to increase in weight of pathogen attached suspended solids that ultimately sink to the bottom. The surfaces of suspended particles offer a nutrient-rich site for bacterial metabolism. This may change their size and shape and consequently their sedimentation rates (Droopo and Orgley, 1974; Wilkinson *et al.*, 1994). Faecal coliforms and other bacteria aggregate into larger particles with the help of the frimbriae or pili so as to hasten settling-out. Wilkinson *et al.*, (1994) have suggested that most bacteria form stable suspensions that can only settle when the system becomes destabilised as in floc formation. They found that sedimentation of bacteria is unlikely to occur under turbulent flow conditions. The role of sedimentation in pathogen removal ranges from 0 to 100% (Maynard *et al.*, 1999). There are several factors, which may affect sedimentation including thermal stratification, turbulence due to wind action. The shape of ponds to achieve effective sedimentation needs to be considered in future experiments. In addition, when bacteria die there is a sloth off and the biofilm fall to the bottom as sediments. This also shows that attachment is a biological process requiring active metabolism in order to take place. In the macrophyte ponds, sedimentation is expected to play a major role in pathogen removal because of the quiescent conditions, which will be created due to the plant cover. The plant cover will reduce turbulence due to wind action.

Most research studies conducted on attachment are focussing on floating macrophytes, because of the visible nature of their roots. Algae being microscopic have not been considered as providing larger surface area for attachment than the macrophytes. The attachment of the pathogens to the algae may enhance die-off, because of generation of harsh environmental conditions on the surface or they could be protected (Mezrioui *et al.*, 1998). Chapter 8 of this thesis will show the role of attachment in faecal bacteria removal in algae and macrophyte based ponds.

Chemical factors and processes

Chemical factors responsible for pathogen removal include pH and dissolved oxygen. The effects of xenobiotic compounds and toxic metals associated with wastewater have not been thoroughly investigated.

pH

The hydrogen ions in aquatic environment are central to many metabolic reactions in microbial cells, including energy generation and ion transport (Mitchell, 1992). This is important in almost all phases of water and wastewater treatment. Aquatic organisms are sensitive to pH changes and biological treatment requires either pH control or monitoring for two reasons. First, discharge requirements generally mandate that pH be between 6.0 and 9.0. Secondly, wastewater with an increased pH may present problems to biological treatment operations that follow. The optimum pH for faecal bacterial growth ranges from 6.5 to 7.5. pH below 4.0 or above 9.5 will reduce the treatment efficiency in terms of BOD removal. Contrarily, levels of pH higher than 9 are effective in pathogen removal (Parhad and Rao, 1974; Pearson *et al.*, 1987b; Curtis *et al.*, 1992a). Frijns and de-Jong (*unpublished*) demonstrated that both high (>8.5) and low (<4.0) pH leads to higher die-off of *E. coli*, whereas a neutral to slightly acidic pH is best for *E. coli* survival.

Considering the removal of faecal coliforms observed at different pH in the dark, Parhad and Rao, (1974) reported a striking increase in the removal of cultured *E. coli* in raw sewage at about pH 9. Mills, (1987) working separately with cultured pond isolates in distilled water, concluded that pH 9 was the major cause for faecal coliform removal. Trousellier *et al.*, (1986) showed that high pH is one of the key causal effects controlling bacteria removal in sewage lagoons. Awuah *et al.*, (2003) also concluded from pH effects studies that high pH (≥ 9) is the major cause of faecal coliform removal in domestic wastewater.

Curtis *et al.*, (1994) also looked at high pH as very important since it not only increased the rate of photo-oxidation but also made the most penetrating wavelength of light bactericidal. They stated that the membranes are the most likely sites of action for exogenously reduced O_2, peroxides, superoxides and hydroxyl radicals. Even after radiations had ceased the sensitivity of faecal coliforms to the high pH persisted. They concluded that membranes were the active sites for exogenously reduced radicals.

Some organisms may behave differently under different pHs. Enterococci were found to be more tolerant to high pH (>9) than to acidic conditions (Awuah *et al.*, 2001). Davies-Colley *et al.*, (1999) also found in their study on waste stabilization ponds that enterococci were also not affected by sunlight when pH was increased from 7.5-9.5 and under low DO levels, low pHs do not have effect on their inactivation. *E. coli* inactivation was however faster at high DO and at high pH.

Causes of pH changes

Carbonate (CO_3^{2-}) and bicarbonate (HCO_3^-) act as the primary buffer for most natural waters. Reactions that produce or consume carbon dioxide (CO_2) may alter the pH temporarily until equilibrium with the atmospheric CO_2 is re-established (Gilmour, 1992). In algal ponds CO_2 is a limiting factor during photosynthesis. In the absence of CO_2, bicarbonates and carbonates in solution are utilised. This alters the balance of the carbonate reactions to produce OH^- and creates alkaline conditions. This phenomenon can alter the pH by as much as 2 to 3 units from neutral (Wetzel, 2001). During the daytime, photosynthesis takes place resulting in increased dissolved oxygen levels. At night there is no photosynthesis, but some of the oxygen produced is used for respiration to produce CO_2, which goes back into the wastewater.

Dissolved oxygen (DO)

The oxygen dissolved in a water body is caused by gas-liquid mass transfer at the surface and subsequent mixing throughout the depth of the water (Baumgartner, 1996). According to this report, rivers can hold only a small amount of DO depending on temperature and turbulence of the water. Ponds hold much less oxygen than flowing water bodies, which are aerated due to frequent agitation by the flowing water. However, in the presence of algae, high levels of oxygen could be produced depending on the algal populations. The oxygen produced by algae in ponds provides the bulk of the oxygen needs of heterotrophic bacteria.

Being facultative anaerobes, faecal coliforms are able to survive under high and low oxygen concentrations. Under anaerobic conditions, faecal coliforms were found to survive for longer periods than under aerobic conditions (Klock, 1971). In their separate studies Klock (1971) and Marais (1974) found that aeration enhanced faecal coliform die-off rates. Kaneko (1997) also found that the removal of polioviruses, bacteriophages and Coxsackie virus B3 is enhanced by aeration. Davies-Colley *et al.*, (1999) observed that F-DNA viruses' inactivation is independent of DO while F-RNA viruses' inactivation increased with increases in DO levels.

Sudden changes in oxygen concentration increased the die-off of *Salmonella typhimurium* strains (Barzily *et al.*, 1991). In aerated water, *Enterococcus faecalis* and *E. coli* were more rapidly inactivated than without aeration in the presence of sunlight (Reed *et al.*, 1988). According to Van Buuren and Hobma (1991), DO concentrations less than 0.5 mg/L did not have any significant effect on the die-off of faecal coliforms. Pearson *et al.*, (1987a) reported that the level of aeration appeared to make little difference to *Clostridium* removal.

Biological factors and processes

Biological factors, which can enhance pathogen removal in wastewater treatment systems, may include attachment, competition for nutrients, predation, allelopathic effects and lysis by viruses. (Attachment has been treated in the previous section).

Effects of the bacterial population
The presence of other bacteria through competition for nutrients and predation mechanisms can have an impact on pathogen removal. Several workers have reported on the importance of nutrient availability to the growth and survival of microorganisms (Atlas and Bartha, 1981; Portier and Palmer, 1989; Mitchell, 1992). High organic loads decrease the rate of pathogen removal in wastewater (Pearson *et al.*, 1987a; Kaneko, 1997; and Almasi and Pescod, 1996). According to Atlas and Bartha, (1981) the intestinal bacteria populations and other pathogens are reduced in numbers and eventually eliminated by competition by the aquatic autochthonous bacteria populations. It has been suggested that nutrient supply and competition for nutrients in the heterotrophic bacteria community are important in determining faecal bacteria die-off rates (Klock, 1971; Wu and Klein, 1976; Legendre *et al.*, 1984).

Some bacteria have been found to reduce the number of viral pathogens. Kim and Unno, (1996) reported that *Bacillus subtilis*, *Pseudomonas aeroginosa* and *Klebsiella pneumoniae* can inactivate polioviruses through predation. However, bacteria are themselves prone to attack by predators. The effects of competition for nutrients of heterotrophic bacteria on pathogen removal in pond systems have not been thoroughly investigated.

Protozoan predation effects
Protozoa prey on organisms such as bacteria, yeasts, algae and other protozoa (Storer *et al.*, 1979). MacCambridge and MacMeekin (1979;1980) have demonstrated the importance of protozoan predation on *E. coli*. Kim and Unno (1996) in their study showed predation of viruses by protozoan to be more efficient than that of bacteria. Some protozoa have been identified to feed on faecal coliform, diptherial, choleral, typhal and streptococcal bacteria species (Enzinger and Cooper, 1976; MacCambridge and MacMeekin, 1979; The conclusions drawn from bacteria predation by protozoa so far reached are that: 1. Slow growing bacteria are eliminated in environments with intense protozoa predation (Atlas and Bartha, 1981; Sinclair and Alexander, 1989); 2. Increasing protozoa populations are also known to positively influence the activated sludge performance by causing an improvement in effluent quality (WPCF, 1990). 3. The protozoa density producing effective removal of bacteria using *Tetrahymena pyriformis* ciliates were $\geq 10^3$ per mL (Curd and Vandyke, 1966).

A major question, which needs to be addressed is what the level of protozoa in pond systems are and what role they play in pathogen removal. Some aspects of these issues have been addressed in chapter 9. Nematodes, rotifers, ciliates, amoebae and bacteriophages are known to feed on bacteria with some specificity (Wilt *et al.*, 1973).

Effects of viruses

Some viruses (bacteriophages including coliphages) attack and lyse bacterial cells to reduce their population (Proctor and Furmann, 1990). Details of pathogen lyses by viruses are not well investigated and more work is needed in this field. The effect of coliphages also needs further investigation. Presence of coliphages in wastewater produced a higher die-off in batch incubation experiments than controls, which were not seeded with cultures of coliphages (Awuah, *unpublished data*).

Effects of macrophytes and algae

Some plants are believed to produce chemical substances, which can either enhance or reduce bacteria survival. Paspalaris and Hodgson (1994) found cold water extracts of the leaves of cypress (*Cupressus macrocarpa*) capable of supporting the growth of *Citrobacter freundii, E. coli, Enterobacter cloacae* and *Salmonella derbyi* while, hot water extracts inhibited all the organisms except *E. cloacae*.

Algae may also have both bactericidal and growth enhancing effects on pathogens. Mezrioui and Oudra (1998) observed that the survival time of *Vibrio cholerae* was extended in an artificial aquatic environment in the presence of algae. In waste stabilization ponds, in the presence of algae (*Chlorella*), *V. cholerae* population reduced more than *E. coli* but cyanobacteria however had the opposite effect (Mezrioui and Oudra, 1998). Similarly, Vela and Guerra (1965) observed that *Salmonella typhi* and *Salmonella paratyphi* grew well in the absence of *Chlorella pyrenoidosa* while *Shigella, Proteus* and streptococci (enterococci) decreased in number when exposed to *Chlorella*. According to Mayo and Noike, (1996), in the presence of the algae *Chlorella vulgaris*, heterotrophic bacteria growth was reduced and this was attributed to competition for glucose with optimum competition occurring at neutral conditions. Competition for glucose declined at high pH. *Chlorella vulgaris* is reported to produce toxins of long chain fatty acids when under stress including high pH (Pratt and Fong, 1940). The algal toxins are clearly selective in pathogen destruction. The populations of algae found in stabilization ponds vary with time (Palmer, 1969). The types of algae in ponds at any given time will affect the indicator organisms differently causing a change in performance if algal toxins or protection are major mechanisms involved in the removal of pathogens in that particular environment. This calls for the identification of algae in ponds.

Retention time

Hydraulic retention time (HRT) in lagoons, which is the period the wastewater will remain in the system before discharge, is of great importance in the die-off/removal of pathogens. If the time of treatment in ponds is extended it enhances die-off. Metcalf and Eddy (2003) have stated that the degree of sedimentation is dependent on the hydraulic retention time. Also, the number of organisms that are removed during wastewater treatment depends on HRT (Oragui *et al.*, 1986) since it allows more time for aggregation, flocculation and settling of suspended particles to which microorganisms including pathogens are attached. The longer the period the higher the chances of pathogens exposed to removal factors such as sunlight, pH, DO and temperature (Rangeby *et al.*, 1996). In domestic wastewater, Feacham *et al.*, (1983) give a range of 30-60 days for pathogenic bacteria removal. For helminths it can even take several months. Grimason *et al.*, (1996a, 1996b) showed that a HRT of 25.3 days is required for the removal of *Ascaris* eggs and *Giardia* cysts in Eldoret, Kenya. However, 40 days will be required to accomplish this in Meze, France. These retention times are longer than the WHO recommended guideline values of 8-10 days to achieve the standard of 1 helminth ovum per litre (WHO, 1989). This may cast doubts on the reliability of the guidelines provided by WHO. Application of the WHO guidelines may produce poorer effluent quality than predicted. In macrophyte-based ponds, HRT may be a critical factor since sunlight mediated actions discussed previously are almost completely absent. The review of the guidelines should consider the technologies in use as well as the country involved.

Depth may increase retention period but may decrease sunlight penetration in algal ponds. Mayo (1989) found that while faecal bacteria reduction in wastewater was observed to increase with solar intensity and hydraulic retention time, it decreased with depth. Agunwamba (1991) and Pearson *et al.*, (1996) also found reduction in pond efficiency with respect to bacteria die-off rate in deeper ponds.

Natural die-off
Natural die-off is one of the most important faecal coliform removal mechanisms. All living things die. This is a natural phenomenon. Generally, faecal coliform decay is observed in natural waters. The die-off is different in different environments. It is generally faster on crops, followed by natural water and then in soil. In soil, pathogens may survive 10 times longer than on crops (Feacham *et al.*, 1983). In wastewater treatment systems, environmental conditions could either increase or reduce the rate of die-off. Bacteria die-off is usually represented with Chick's Law equation as $Ct/Co=e^{-kt}$, where k (often expressed as k^{-d}) is the die-off rate, t is the retention time, Ct is the concentration of bacteria (number per volume) at any time t and Co is the initial concentration. The suggested WHO value to be used for designing WSP is 1.07 (Johansson *et al.*, 1996).

Saqqar and Pescod (1992) observed that k increased with increase in temperature, solar radiation and pH. Correlation analysis showed that k is more responsive to temperature than solar radiation and pH. Mara *et al.*, (1986) found the die-off rate in primary ponds in Kenya to be high in anaerobic ponds and low in maturation ponds. This however, changed when algal bloom of *Scenedesmus quadricauda* occurred in the pond and created a pH around 10. At this pH no coliform was detected. Studies in Portugal showed that temperature, precipitation and sunlight have an effect on bacterial die-off in waste stabilization ponds (Nascimento *et al.*, 1991). It was observed that for faecal streptococci, faecal coliforms, *Clostridium perfringens*, *Pseudomonas aeruginosa*, and heterotrophic bacteria, the die-off rate varied not only in the same pond but also from pond to pond. Influence of climatic factors was strongest in anaerobic ponds and least in maturation ponds.

The die-off rate is higher in wastewater with a low organic load. The k value in UASB ponds is much higher than that in anaerobic ponds (Van Haandel and Lettinga, 1994; Catunda *et al.*, 1996). Others, including Dewedar and Bahgat (1995), have found that in field experiments, faecal coliform die-off rate (k^{-d}) increases after stationary phase 3 to 5 days later. Troussellier *et al.*, (1986) found that when the faecal coliform concentrations were high, the removal rate for faecal coliforms was high despite an inadequate treatment period in the lagoons. They concluded that other factors therefore must be involved in the removal mechanism. Mubanga (2002) also observed that when *E. coli* concentrations are high they tend to die faster than at lower concentrations. The die-off rate, which is generally, based on Chick's law uses exponential equations. It must be noted that not all bacterial die off may follow this equation. The natural die-off occurs as a result of the environmental conditions and the long retention periods.

Concluding remarks

The pathogen removal mechanisms in pond systems focus on maturation ponds where algae mediated reactions produce high DOs and high pH. The mechanisms in facultative and anaerobic ponds are not clear in the literature. It has however been demonstrated throughout the review that sunlight is very important in pathogen removal.

Other mechanisms involved in pathogen removal in pond systems have not been studied extensively as sunlight and pH. In macrophyte ponds the literature is scarce on the subject matter. Since sunlight is virtually absent in macrophyte ponds, the removal of pathogens should be enhanced through other mechanisms, which promotes the destruction of pathogens in the absence of sunlight. The complex nature of microorganisms and the environment in which they operate will not make the design of experiments easy. Results may also be difficult to explain in some cases. Literature on helminths, viruses and protozoa is scarce. Not much is being done in this field.

All mechanisms discussed are important depending on the pathogen under investigation. The extent of each important mechanism must be examined and incorporated in the design of wastewater treatment for optimisation. In this review, time may be the critical factor, which must be looked at since environmental conditions need time to act and to promote natural die-off, which may be the key factor in waste stabilization ponds.

In algal ponds, sunlight triggered mechanisms associated with pH, dissolved oxygen and sedimentation may be the key factors in pathogen removal. Bacteria predation by protozoa may play a significant role if protozoa population levels are high. Attachment may play a significant role in pathogen removal. The pathogens may get attached to algae and get exposed to the harsh conditions created in the ponds around the algal cells.

In macrophyte ponds, sedimentation, attachment, long retention periods and predation may be the key factors for pathogen removal.

More research work should be conducted on the removal mechanisms for viruses, protozoa and helminth eggs. Studies into the mechanisms other than sunlight triggered ones involved in pathogen removal especially in macrophyte ponds should be encouraged in the wake of resource recovery options in wastewater treatment of today.

Scope and content of PhD thesis

This PhD thesis concentrates on some of the issues raised by looking at the performance of macrophytes and algal waste stabilization ponds and their environmental conditions. In addition, the effect of pH, attachment and protozoa grazing on faecal bacteria removal in waste stabilization ponds has been studied to some extent. The study was however limited to short incubation studies on batch scale, bench and pilot scale continuous flow systems. Chapter 1 which is the introduction also provides an overview of the pathogen removal mechanisms in pond systems.

Chapter 2 looks at the performance and environmental conditions in macrophyte and algal ponds on batch-scale. The environmental conditions in continuous flow systems and effect of pH in the removal of enterococci are examined in chapter 3. In chapter 4, the environmental conditions in bench-scale continuous flow systems are examined in detail using diluted sewage and the effect of pH on faecal coliform removal is evaluated. The effect of fluctuations in pH (which are common in algal ponds) on removal of *E. coli*, coliforms, *Salmonella* and other enterobacteria are also studied in chapter 5.

Chapter 6 examines the performance of macrophyte and algal ponds using diluted sewage. Emphasis is placed on faecal coliform removal and population at different levels in both macrophyte and algal waste stabilization ponds.

Chapter 7 looks at the performance of organic and nutrient loads removal and faecal bacteria removal using raw sewage (medium strength sewage). Emphases were put on faecal bacteria removal using *E. coli*, coliforms, *Salmonella*, other enterobacteria and enterococci.

In Chapter 8, the possible role of attachment in the removal of faecal bacteria is studied both in a batch scale and in a continuous flow system.

In chapter 9 the feasibility of predation by protozoa on faecal bacteria was assessed in macrophyte and algal ponds by looking at protozoa population profiles. Batch-scale experiments on removal of faecal bacteria in the presence and absence of protozoa were also carried out.

References

Actis LA, Rhodes ER and Tomaras AP (2003). Genetic and molecular characterization of a dental pathogen using genome-wide approaches. *Adv Dent Res* **17**:95-99.

Agunwamba JC (1991). Simplified optimal design of wastewater stabilization pond. *Wat. Air, Soil, Pollu.* 59(3/4), 299-309.

Almasi A and Pescod MB (1996). Pathogen removal mechanisms in anoxic waste stabilization ponds. *Wat. Sci. Tech.* **33**(7), 133 – 140.

Atlas RM and Bartha R (1981). Microbial ecology; Fundamentals and applications. Addison-Wesley Publishers Co. London, 560p.

Awuah E, Anohene F, Asante K, Lubberding HJ and Gijzen HJ (2001). Environmental conditions and pathogen removal in macrophyte and algal-based domestic wastewater treatment systems. *Wat. Sci. Tech.* 44 (6),11-18.

Awuah E, Kuffour AR, Lubberding HJ and Gijzen HJ (2002). Characterization and management of domestic wastewater in two suburbs of Kumasi. In: *Proceedings of Water and Health Conference* Ottawa, Canada, September 2002. 475p.

Awuah E, Boateng J, Lubberding HJ and Gijzen HJ (2003). Physico-chemical parameters and their effects on pathogens in domestic wastewater In: *KNUST SERR 2 Proceedings*. Elmina, September, 2002. 261p.

Barzily A and Kott Y (1991). Survival of pathogenic bacteria in an adverse environment. *Wat. Sci. Tech.* **24**(2), 395-400.

Baumgartner DJ (1996). Surface water pollution. Pollution Science. Pepper JL, Gerba PC and Brusseau ML (eds). Academic Press Inc. Canada, pp. 189 – 220.

Bitton G (1994). Wastewater microbiology. John Wiley and Sons New York,. 478p.

Bos AA, Gijzen HJ, Hilderink H, Moussa M, de Ruyter E, Niessen L (2005) Health benefits versus costs of water supply and sanitation. *Water21*, October 2005: 31—35.

Burkhardt III W, Kern R, Calci W, Watkins D, Rippey SR and Chirtel SJ (2000). Inactivation of indicator organisms in estuarine waters. *Wat. Res.* **34**, 2207-2214.

Byamukama D, Kansiime F, Mach RL and Farnleitner H (2000). Determination of *Escherichia coli* contamination with chromocult coliform agar showed a high level of discrimination efficiency for differing faecal pollution levels in tropical waters of Kampala, Uganda. *Appl. Environ. Microbiol.* **66,** 864-868.

Byappanahalli M and Fujioka R (2004). Indigenous soil bacteria and low moisture may limit but allow faecal bacteria to multiply and become a minor population in topical soil. *Wat. Sci. Tech.* **50**(1), 27-32.

Catunda FC, Van Haandel AC, Mara DD, Pearson HW and Silva SA (1996). Improved performance and increased applicability of waste stabilization pond by pre-treatment in a UASB reactor. *Wat. Sci. Tech.* **33**(7), 147-157.

Cavari B and Bergstein T (1996). Factors affecting survival of pathogen indicators of pollution of freshwaters. Israel Oceanographic & Limnological Research. Tiberias, Israel. pp. 412-415.

Cetinkaya Y, Falk P and Mayhall CG (2000). Vancomycin-resistant enterococci. *Clinical Microbiol. Rev.*, **13**, 686-707.

Cosgrove WJ and Rijsberman FR (2000). World Water vision. Making water everybody's business. The world water council, Earthcan Publishers Ltd., UK, pp. 108.

Crites R and Tchobanoglous G (1998). Small and decentralised wastewater management systems. WCB Mc Graw-Hill pp. 80-86.

Curds CR and Vandyke JM (1966). The feeding habits and growth rates of some fresh ciliates found in activated sludge plants. *J. Appl. Ecol.* **3**, 127-137.

Curtis TP (1990). The mechanisms of removal of faecal coliforms from waste stabilization ponds. PhD. thesis, University of Leeds.

Curtis TP, Mara DD, and Silva SA (1992a). Influence of pH, oxygen, and humic substances on ability of sunlight to damage faecal coliforms in waste stabilization pond water. *Appl. Environ. Microbiol.* **58**, 1335-1343.

Curtis TP, Mara DD, and Silva S A (1992b). The effect of sunlight on faecal coliforms in ponds: implications for research and design. *Wat. Sci. Tech.* **26** (7/8), 1729-1738.

Curtis TP, Mara DD, Dixo NGH and Silva SA (1994) Light penetration in waste stabilization ponds. *Wat. Res.* **28**, 1031-1038.

Davies-Colley RJ, Donnison AM, Speed DJ, Ross CM and Nagels JW (1997). Sunlight wavelengths inactivating faecal indicator microorganism in waste stabilization ponds. *Wat. Sci. Tech.* **35**(5), 219-225.

Davies-Colley RJ, Donnison AM, Speed DJ, Ross CM and Nagels JW (1999). *Wat. Res.* **33**, 1220-1230.

Dewedar A and Bahgat M (1995). Fate of faecal coliform bacteria in a wastewater retention reservoir containing *Lemna gibba* L. *Wat. Res.* **29**, 2598 – 2600.

Droopo M and Jannasch H (1980). Advantages of Aquatic Microbiology. Academic press, New York. 23-25.

Droopo M and Orgley E (1974). Flocculation of suspended sediment in rivers of south eastern Canada. *Nat. Wat. Res. Inst.* **28**, 1799-1809.

Enzinger RM and Cooper RC (1976). Role of bacteria and protozoa in the removal of Escherichia coli from estuarine waters. *Appl. Environ. Microbiol.* **31**, 758-763,

Feachem RG, Bradley DJ, Garelick H and Mara DD (1983). Sanitation and Disease: Health Aspects of Excreta and wastewater management. Published for the World Bank by John Wiley and Sons, U.K.

Fletcher MM (1996). Bacteria Adhesion: molecular and ecological diversity. Wiley-Liss Inc., New York. 371p.

Gannon JJ, Buse K and Schillinger J (1983). Faecal coliform disappearance in a river impoundment. *Wat. Res.* **17**, 1595 – 1601.

Garcia M, Bécarès E, Morris R, Grabow WOK and Joefre J. (1997). Bacteria removal in three pilot scale wastewater treatment systems for rural areas. *Wat. Sci. Tech.* **35** (11-12), 197-200

Gersberg R, Brenner R, Lyon S and Elkins B (1987). Survival of bacteria and viruses in municipal wastewater applied to artificial wetlands. In: *Aquatic plants for wastewater treatment using artificial wetlands*. Reddy K and Smith W (eds.); Magnolia Publishing, Orlando Florida, pp. 237-245.

Gibbs RA, Hu CJ, Ho GE and Unkovich I. (1997). Regrowth of faecal coliforms and salmonellae in stored biosolids and soil amended with bio/solids. *Wat. Sci. Tech.* **35**(11/12), 269 – 175.

Gijzen HJ and Khonker M (1997). An overview of ecology, physiology cultivation and application of duckweed. Literature review. Report No. 0896. Duckweed research project. Bangladesh, pp. 69

Gijzen HJ (2001). Anaerobes, aerobes and phototrophs – A winning team for wastewater management. *Wat. Sci Tech* **44**:8, 1213-132.

Gilmour CC (1992). Effects of acid deposition on microbial processes in natural waters. In: Mitchell, R (ed). *Environmental microbiology*. Wiley-Liss Inc. New York, pp. 33-57.

Gray NF (2004). Biology of wastewater treatment. Second edition. Imperial College press London, 1421p.

Greenberg AE, Clesceri LS, and Eaton AD (2003). Standard methods for the examination of water and wastewater. APHA/AWWA. Water Environmental Fed. Washington D.C.

Grimason AM, Smith HV, Young G and Thitai WN (1996a). Occurrence and removal of *Ascaris sp.* Ova by waste stabilization ponds in Kenya. *Wat. Sci. Tech.* **33**(7), 75-82.

Grimason AM, Wiandt S, Baleux B, Thitai WN, Bontoux J and Smith HV (1996b). Occurrence and removal of *Giardia sp.* cysts by Kenyan and French waste stabilization pond systems. *Wat. Sci. Tech.* **33**(7), 83-89.

Haag WR, Hoigne J, Gassman E and Brunn A (1986). Singlet oxygen in surface water, Part III. Photochemical formation and steady-state concentrations in various types of water. *Chemosphere* **13**, 641 – 650.

Harm W (1968). Dark repair of photorepairable UV lessions in *Escherichia coli. Mutat. Res.* **6**, 25-35.

Health News, (2004).:http://www.intelihealth.com/IH/ihtIH/EMIHC000/333/29758.html http://water.sesep.drexel.edu/outbreaks/Sydney_5/r5vol1.htm

Jagger J (1985). Solar-UV radiations on Living cells. Praeger, New York, pp. 174-176.

Johansson P, Penrup M and Rangeby M (1996). Low cost up-grading of an oversized waste stabilization pond system in Mindero, Cape Verde. *Wat. Sci. Tech.* **33**(7), 99-106.

Kadlec R and Hammer D (1982) Pollutant transport in wetlands. *Environ. Progr.* **11**, 206-211.

Kaneko M. (1997). Virus removal by domestic wastewater treatment system *Wat Sci. Tech.* **35** (11/12), 187-195.

Kim TD and Unno H (1996). The role of microbes in the removal and inactivation of viruses in a biological wastewater treatment system. *Wat. Sci. Tech.* **33**(10/11), 243-250.

Klock JW (1971). Survival of faecal coliforms in wastewater. *WPCF* 50, 20171-20183.

Klug M and Reddy C (1984) Current perspectives in microbial ecology. Proceedings of the 3rd International Symposium on Microbial Ecology. Michigan State University. *American Soc. Microbiol.* Washington D.C. pp. 130-138.

Konig A (1984). Ecophysiological studies on some algae and bacteria of waste stabilization ponds PhD Thesis, University of Liverpool. United Kingdom.

Landsdell M (1987). The development of lagoons in Venezuela. *Wat. Sci. Tech.* **19**(12), 55-60.

Legendre P, Baleux B and Troussellier M (1984). Dynamics of pollution-indicator and heterotrophic bacteria in sewage treatment lagoons. *Appl. Environ. Microbiol.* **48**, 486-593.

Len Y Wen-Shi C and Mong-Na LH (2000). Natural disinfection of wastewater in marine outfields. *Wat.Res.* **34,** 743-750.

Lesgne, J Baleux B Boussaid A and Hassan L (1991). Dynamics of non01 *Vibrio cholerae* in experimental sewage stabilization ponds under Mediterranean conditions. *Wat. Sci. Tech.* **24**(2), 387-390

Liltved H and Landfold B (1996). Influence of inactivation of survival of ultraviolet irradiated fish pathogenic bacteria. *Wat. Res.* **30**, 1109-1114.

Liltved H and Landfold B (2000). Effects of high intensity light on ultraviolet –irradiated and non irradiated fish pathogenic bacteria. *Wat. Res.* **34**, 81-486.

Mandi L, Ouzzani N, Bouhoum K and Boussaid A (1993). Wastewater treatment by stabilization ponds with and without macrophytes under arid climate *Wat. Sci. Tech.* **28** (10), 177-181.

Mara DD and Pearson HW (1986). Artificial freshwater environments: waste stabilization ponds. In: Schoernborn W (ed) *Biotechnology*, Weinheim, Germany: VCH Verlagsgesellschaft, pp.177-206.

Marais GVR (1974). Faecal bacterial kinetics in stabilization ponds. *Environ. Eng. Div.* ASCE **100**, 119-139.

Marshall K (1973). Mechanism of bacteria adhesion at solid-water interfaces. University of New South Wales, Kensington, New South Wales publishers. 133-142.

Maynard HE, Ouki SK, and Williams SC (1999). Tertiary Lagoons: A review of removal mechanisms and performance. *Wat. Res.* **33**, 1 – 13.

Mayo AW (1989). Effect of pond depth on bacterial mortality rate. *J. Environ. Eng.* **115**, 964-977.

Mayo AW and Noike T (1996). Effects of temperature and pH on the growth of heterotrophic bacteria in waste stabilization ponds. *Wat. Res.* **30**, 447-455.

McCambridge J and MacMeekin T (1979). Protozoan predation of Escherichia coli in estuarine waters. *Wat. Res.*, **13**, 659-663.

McCambridge J and MacMeekin T (1980). Relative effects of bacterial and protozoa predators on survival of E. coli in estuarine water samples. *Appl. Environ. Microbiol.* **40**, 907-911

Meckes M.C (1982). Effect of UV light disinfection on antibiotic resistant coliforms in wastewater effluents. *Appl. Environ. Microbiol.* **43**, 371-377

Metcalf and Eddy (2003). Wastewater Engineering: Treatment, disposal and reuse. McGraw Hill Inc. New York.

Mezrioui N, Baleaux B and Troussellier M (1995). A microcosm study of the survival of *Escherichia coli* and *Salmonella typhimurium* in brackish water. *Wat. Res.* **29**, 459-465.

Mezrioui NE and Oudra B (1998). Dynamics of picoplankton and microplankton flora in the experimental wastewater stabilization ponds of the arid region of Marrakech, Morocco and cyanobacteria effects on *Escherichia coli* and *Vibrio cholerae* survival. In: *Wastewater treatment with algae,* Wong Y-S and Tam NFY (eds). Springer-Verlag and Landes Bioscience, pp. 165-188, p. 234.

Mills SW (1987). Sewage treatment in waste stabilization ponds: physiological studies on the microalgal and faecal coliform populations. PhD. Thesis, University of Liverpool, U.K.

Mitchell R (1992). Environmental microbiology. Wiley-liss Inc. New York. 411p.

Moeller JR and Calkins J (1980). Bactericidal agents in wastewater lagoons and lagoon design. *WPC F* **52**, 2443-2451.

Moore BE (1993). Survival of human immunodeficiency virus (HIV), HIV-infected lymphocytes, and poliovirus in water. *Appl. Environ. Microbiol.* **59**, 1437–1443.

Mubanga J 2002. The potential contribution of attachment to the removal of pathogens from wastewater treatment plants MSc. Thesis, WERM 0215, IHE, Delft, Netherlands

Nascimento MJ, Oliveira JS and Mexia JT (1991). Contribution for the study of new pathogenic indicators removal from waste stabilization pond in Portugal. *Wat. Sci. Tech.* **24**(2), 381-386.

Nhapi I and Gijzen HJ (2005). A 3-step strategic approach to sustainable wastewater management. *Water SA*, **31**, 133-140.

Oragui JI, Curtis TP, Silva SA and Mara DD (1986). The Removal of excreted bacteria and viruses in deep waste stabilization ponds in Northeast Brazil. *Wat. Sci. Tech.* **18** (7/8), 31-35.

Orlob GT (1956). Viability of sewage bacteria in seawater. *Sewage and Industrial Wastes* **28**, 1147-1167

Otterpohl R, Abold A and Oldenburg M (1998) *Differentiating management of Water Resources and Waste in Urban Areas*, Proceedings of the internet conference on integrated biosystems Lübeck, Germany: Internet Site. http://www.soc.titech.ac.jp/uem/waste/oldenburg.html

Palmer CM (1969). A composite rating of algae tolerating organic pollution. *J. Phycol.* **5**, 78-82.

Parhad NM and Rao NU (1972). The effect of algal growth on the survival of *E coli* in sewage. *Indian J. Environ. Health.* **14**, 131-139.

Parhad NM and Rao NU (1974). Effect of pH on survival of *E. coli. J. Wat. Pollut. Contr. Fed.* **46**, 149-161.

Paspalaris P and Hodgson B (1994). The role of Cypress leaves in a holding reservoir. *Wat. Res.* **28**, 2147-2151.

Pearson HW, Mara DD, Cawley LR, Arridge HM. and Silva SA (1996). The performance of an innovative tropical experimental waste stabilization pond system operating at high organic loadings. *Wat. Sci. Tech.* **33**(7), 63 – 73.

Pearson HW, Mara DD, Mills SW and Smallman DJ (1987a). Factors determining algal populations in waste stabilization ponds and influence on algae pond performance. *Wat. Sci. Tech.* **19**(12), 131-140.

Pearson HW, Mara DD, Mills SW and Smallman DJ (1987b). Physiochemical parameters influencing faecal bacteria survival in waste stabilization ponds. *Wat. Sci. Tech.*, **19**(12). 145-152.

Polprasert C, Dissanayake MG and Thanh NC (1983) Bacterial die-off kinetics in waste stabilization ponds. *J. Water Pollut. Control Fed.* **55**, 285-296.

Polprasert C (1996). Organic waste recycling. John Wiley and Sons 426p. New York.

Portier R and Palmer S (1989). Wetlands microbiology: Form, function, process. In: *Constructed wetlands for wastewater treatment: Municipal, industrial and agricultural.* Hammer DA (ed). Lewis publishers Chelsea, Michigan, pp. 89-106.

Pons M, Spanjers H, Baetens D, Nowak O, Gillot S, Nouwen J and Schuttinga N (2004). European water management on line. EWA, pp.10. Internet site: http://www.ewaonline.de/journal/2004_04.pdf#search='wastewater%20characteristics

Pratt R and Fong J (1940). Studies on *Chlorella vulgaris*: Further evidence that *Chlorella* cells form a growth inhibiting substance. *Am. J. Bot.* **27**, 431-436.

Proctor LM and Fuhrman JA (1990). Viral mortality of marine bacteria and cyanobacteria. *Nature,* **343**, 60-62.

Rangeby M, Johansson P and Pernrup M (1996). Removal of faecal coliforms in a waste stabilization pond system in Mindelo, Cape Verde. *Wat. Sci. Tech.* **34** (11), 149 – 157.

Reed SC, Middlebrooks EJ and Crites RH (1988). Natural systems for waste management and treatment. McGraw-Hill, USA

Rheinheimer G (1992), "Aquatic microbiology", 4th Edition. John Wiley and Sons, Chichester, 184p.

Rozac DB and Colwell RR (1987) Survival strategies of Bacteria. *Microbial Rev.* **51**, 365-378.

Salvato JA (1992) Environmental Engineering and Sanitation. 4th edition John Wiley and Sons, New York 643pp.

Saqqar MM and Pescod MB (1992). Modelling coliform reduction in waste stabilization ponds. *Wat. Sci. Tech.* **26**(7/8), 1667-1677.

Sinclair JL and Alexander M (1989). Effect of protozoa predation on relative abundance of fast and slow growing bacteria. *Can. J. Microbiol.* **35**, 578-582.

Storer TI, Usinger RL, Stebbins RC and Nybakken JW (1979). General Zoology, Sixth Edition. McGraw-Hill Book Company, New York.: i-ix, 1-902.

Tortora GJ, Funke BR and Case CL (2003). Microbiology. An introduction. 8th Edition. Benjamin/Cumming Press California, 944p.

Troussellier M, Legendre P and Baleux B (1986). Modelling of the evolution of bacterial densities in an eutrophic ecosystem (sewage lagoons). *Microbial Ecol.* **12**, 355-379.

USEPA (1988). Design manual for constructed wetland and aquatic plant systems for municipal wastewater treatment, *USEPA 625/ 1-88 / 022*

Van Buuren JCL and Hobma S (1991). The faecal coliform removal rate at post treatment of anaerobically pre-treated domestic wastewater. Department of Environ. Tech. Agricultural University, Wageningen *Unpublished*

Van der Steen P, Brenner A. Shabtai Y, Oron G. (2000). Improved faecal coliform decay in integrated duckweed and algal ponds. *Wat. Sci. Tech.* **42** (10/11), 363-370.

Van Haandel AC, and Lettinga G (1994). Anaerobic sewage treatment. John Wiley and Sons New York,. 226p.

Vela GR and Guerra GN (1965). On the nature of mixed cultures of *Chlorella pyrenoidosa* TX 7110-5 and various bacteria. *J. Gen Microbiology* **42**, 123-131.

Vincent G (1994). Use of artificial wetlands for the treatment of recreational wastewater. *Wat. Sci. Tech.* **29** (4), 67-70.

Von Sperling M. (1996). Comparison among the most frequently used systems for wastewater treatment in developing countries. *Wat. Sci. Tech.* **33**(3), 59-76.

Wang B (1991). Ecological waste treatment and utilization system: low-cost energy saving/generating, resources recoverable technology for pollution control in China. *Wat. Sci. Tech.* **24** (1), 8-19.

WPCF (1990) Wastewater biology. The microlife. WPCF Press Alexandria Virginia, pp. 196.

Wetzel RG (2001). Limnology of lake and river ecosystems. 3rd Edition Academic press. San Diego, 429p.

Whitlam GC and Codd GA (1986). Damaging effects of light on microorganisms, *Spec. Publ. Soc. Gen. Microbiol.* **17**, 129 – 169.

Wilkinson J, Jenkin A, Wyer M and Kay D (1994). Modelling of faecal coliforms dynamics in streams and rivers. *Wat. Res.* **3**, 847-855.

Wilt GR, Joshi MM and Metcalf J (1973). Studies on bacteria and nematodes. *Alabama Water Resources Research Institute Bulletin* **10**, 89p.

WHO (1989). Health guidelines for the use of wastewater in agriculture and aquaculture (*Technical Report # 778*). WHO, Geneva.

WHO (2005). Water for life: making it happen. Geneva, Switzerland, World Health Organization. 38p.

Wu S and Klein DA (1976). Starvation effect on *Escherichia coli* and aquatic bacteria to nutrient addition and secondary warming stresses. *J. Appl. Environ. Microbiol.* **31**, 216-220.

Zepp RG (1988). Environmental photoprocesses involving natural organic matter. In: *Humic substances and their role in environments.* Frimmel FH and Christian RF (eds) Wiley Publishers, New York, pp. 193-214.

Zimmo O (2003). Nitrogen transformations and removal mechanisms in algal and duckweed waste stabilization ponds. PhD Thesis. UNESCO-IHE, Delft Netherlands, 131p.

Chapter Two

Environmental Conditions and Enterococci Removal in Macrophyte and Algal-Based Domestic Wastewater Treatment Systems

Adapted from

Awuah E, Anohene F, Asante K, Lubberding HJ and Gijzen HJ (2001).

Environmental conditions and pathogen removal in macrophyte and algal-based domestic wastewater treatment systems

Water Science and Technology **44**(6), 11-18

Environmental conditions and enterococci removal in macrophyte and algal-based domestic wastewater treatment systems

Abstract

The environmental conditions and faecal enterococci removal in batch-scale water lettuce (*Pistia stratiotes),* duckweed (*Lemna paucicostata*) and algal ponds were determined over a period of 29 days under tropical conditions. Batch-scale incubation experiments were conducted in 4.5L plastic containers. A control of raw sewage stored under dark conditions was included. Environmental conditions such as temperature, pH, and DO, heterotrophic bacteria and enterococci populations were monitored four times a week at 8, 12 and 20 GMT. BOD and COD were monitored once a week for five weeks. Average temperatures within the systems ranged between 28.3°C in the raw wastewater to 30.6°C in the algal ponds. Low levels of pH 4.5 and DO levels of 3mg/L were recorded in the water lettuce ponds. The duckweed ponds had neutral pH and DO of 6mg/L. The raw wastewater under darkness had neutral pH and low DO concentrations. High pH levels around 10.5 and DO of about 20mg/L were observed in the algal ponds in the afternoons. All pond systems performed equally well in enterococci removal and BOD reduction. The enterococci population decreased from 1.2×10^4 /100mL to values < 100/100mL in all treatment systems. The BOD decreased from 130mg/L to 5.0, 7.5, 10 and 15mg/L in the duckweed, water lettuce, raw wastewater and algal treatment systems respectively.

Keywords Water lettuce; duckweed; enterococci; macrophytes; wastewater

Introduction

Wastewater treatment plants in many developing countries have broken down due to introduction of advanced technologies, high operation and maintenance cost, and lack of skilled personnel, logistics and incentives. As a solution to the above problems, natural purification systems such as waste stabilization ponds have also been introduced. Unfortunately, these have also been neglected and poorly maintained including recently constructed ones that are under operation in Ghana (Awuah *et al.*, 1996; Salifu, 1996; Awuah *et al.*, 2002a). This calls for resource recovery technologies to generate income as an incentive to wastewater treatment in Ghana. Reliance on floating macrophyte (aquatic plants) is gaining recognition world-wide because of the ability of these plants to remove nutrients from wastewater and the potential use of the plants for aquaculture and poultry (Gijzen and Khonker, 1997; Gijzen and Ikramullah, 1999; Nhapi, 2003). The removal of pathogens is of utmost importance since enteric disease is one of the major causes of death in children in developing countries (UNICEF/UNEP, 1990). In Ghana, enteric disease is second to malaria in morbidity (Ministry of Health, 1996). There are, however, conflicting reports on the efficiency of pathogen removal in macrophyte and algal ponds. Studies with indicator organisms indicate that the removal of pathogens is more efficient in algal ponds than in macrophyte ponds (Gijzen and Khonker, 1997). Other reports indicate that macrophyte ponds are more efficient in the removal of pathogens than algal ponds (Mandi *et al.*, 1993; Garcia and Bécarès, 1997). It must be noted that these experiments were conducted in different parts of the world with different climates and wastewater characteristics. The removal of pathogens will depend on the environmental conditions within the ponds. The treatment processes in natural systems are very complex and are influenced by many factors such as hydraulic and hydrological characteristics, patterns of flow, retention time, biological characteristics like species composition, plant life forms, root mat structure, microbial activity, limno-chemical features and nutrient mass balances (Denny, 1997). The efficiency of pathogen removal in algal ponds is well documented. These have been attributed to sunlight induced physico-chemical parameters. Sunlight through photosynthesis increases the pH and dissolved oxygen (DO) levels in the wastewater and this has been documented as the main factors influencing pathogen removal (Parhad and Rao, 1972; Curtis, 1990; Curtis et al., 1992). In spite of this, other researchers have also shown that sunlight may not be the only mechanism involved in pathogen removal (Almasi and Pescod, 1996). In macrophyte ponds, the cover of floating aquatic plants eliminates the direct effect of sunlight. Hence the conditions within such systems, which determine the removal of pathogens needs to be investigated. The aim of the overall research is to determine the pathogen removal mechanisms in algal and macrophyte ponds. The objective of this study was therefore to make a comparative study of the environmental conditions within macrophyte and algal ponds and their general performance, particularly enterococci removal.

Materials and Methods

Location

The study was carried out at the Kwame Nkrumah University of Science and Technology (KNUST) at Kumasi (Ghana). Kumasi is located in the tropical forest belt of Ghana between latitude 6.40^0 and 6.35^0 N and longitude 1.30 and 1.35 W. It is at 250-399m above sea level. The average temperature is 27.8^0C and the annual rainfall is 1300mm.

Macrophyte Selection

Water lettuce (*Pistia stratiotes*) and duckweed (*Lemna paucicostata*) were selected primarily on availability and economic uses. Even though the use of water lettuce for wastewater treatment has not been studied in detail its treatment efficiency may be comparable to that of water hyacinth (Brix and Schierup, 1989). Duckweed was found in many of the ponds around the University. Water lettuce was also found in some streams, although it was rarer in Kumasi than duckweed. Water lettuce is used as a pig feed and for curing asthma in Ghana. Duckweed has no economic importance in Ghana now but chickens have been seen eating it.

Aquatic Plant Cultures

Water lettuce was cultured in tap water with NPK (15+15+15 by weight) fertilizer at a concentration of 0.05 mg/L under sunlight conditions. Duckweed was cultured in sterilized sewage in the laboratory. Duckweed was exposed to ambient environmental conditions for at least three days prior to the introduction in the sewage. Algae were introduced into the algal pond by natural colonization.

Experimental Set Up

A wooden box of size 1.2m x 0.8m x 0.2m was constructed. This was placed on a table at a height of 1m in the open. The set up was roofed with transparent polythene sheet to protect the experiments from rain. Nine 4.5L white plastic containers were arranged in the box. The box was then filled with sand, which was kept moist to prevent sunlight penetration from the sides and to regulate temperature respectively.

Fresh sewage from KNUST wastewater treatment plant was digested anaerobically in the laboratory for two days. Four litres of the digested sewage was poured into each container. The water lettuce, duckweed and algae treatments were arranged in randomized block design and exposed to sunlight. A control consisting of raw sewage without macrophyte or algae was kept in the dark. Twenty-five grams fresh weight of water lettuce with an average of 20 roots was put in each water lettuce pond, while duckweed ponds were started with 10g fresh weight. Water lettuce was harvested once in a week and duckweed was harvested twice in a week.

Monitoring frequencies and laboratory analyses

The environmental conditions of the ponds were monitored 3 times a day, four times a week for 5 weeks. Heterotrophic and enterococci bacteria were determined from samples taken at a frequency of once a week and three times a day.
Biochemical oxygen demand ((BOD) and chemical oxygen demand (COD) were measured once a week for five weeks. Ammonia, nitrate, nitrite were measured at the initial and final stages of the study.

Faecal enterococci were used as indicator organisms. Enterococci were determined by the pour plate method on Slanetz Bartley agar after incubation at 44.5^0C for 24 hours (Niemi and Ahtiainen, 1995). Heterotrophic bacteria were determined by using total bacteria plate count media after incubation at 37^0C for 24 hours.

Analytical methods in Greenberg et al., (1992) were adopted in the laboratory determinations of BOD, COD, ammonia, nitrate and nitrites. Temperature, pH and DO were measured in situ with portable electronic probe Microprocessor Oxi /pH/mV 323/325 meters.

Results

Environmental Conditions

There were variations in environmental conditions between 8GMT and 20GMT readings for temperature, pH and DO in all pond systems (Figures 1-3). The temperature in the water lettuce, duckweed and algal system followed a similar pattern during the whole experiment with temperatures between 27 and 32^0C. The temperature in the control, which was kept in the dark, remained constant at lower values between 27 and 28^0C (Figure 1).

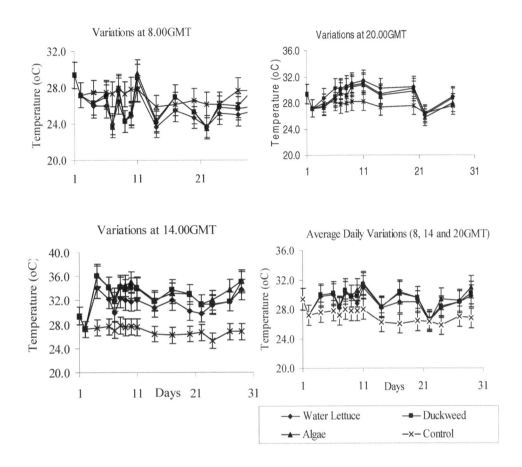

Figure 1 Temperature variations in macrophyte and algal ponds

Water lettuce ponds decreased in pH over time and reached the lowest level (pH 4.5) on the 20th day after which the pH started rising gradually. The pH in the raw sewage and the duckweed system stayed neutral 7-7.8 and turned slightly alkaline (8.6) towards the last few days when algae colonization set in due to poor growth. The alkaline conditions in the algal ponds rose to its peak in the first week and remained high till the end of the experiment. The algal ponds fluctuated between neutral and alkaline conditions at 8GMT and 14GMT respectively. pH values reached 10.5 at 14GMT (Figure 2).

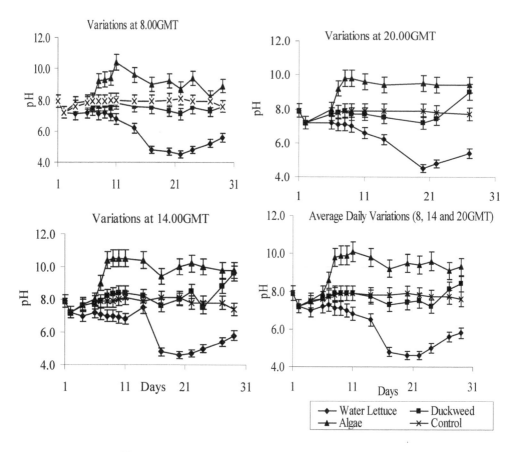

Figure 2 pH variations in macrophyte and algal ponds

There was a big difference between the DO patterns in all four systems. A minimal increase of up to 1 mg/L after two weeks was seen in the (raw sewage control) due to air diffusion. The water lettuce and duckweed brought more O_2 into the water due to photosynthesis, showing levels of up to 3 and 6 mg/L, respectively. Dissolved oxygen concentrations in the algal ponds, which were beyond saturation levels had readings of up to about 20mg/L in the afternoons (Figure 3).

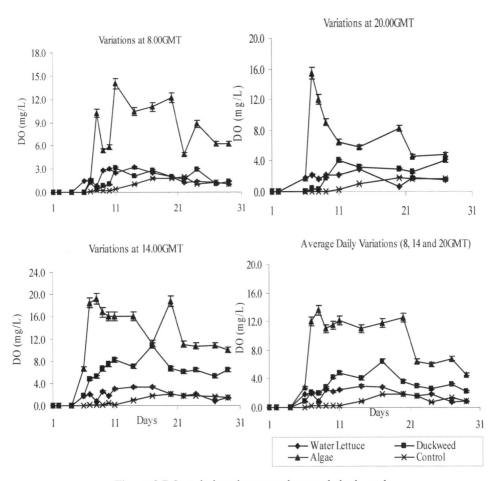

Figure 3 DO variations in macrophyte and algal ponds

Heterotrophic and enterococci bacteria removal

Heterotrophic and enterococci bacteria started decreasing from the onset to the end of the experiment in all the pond systems. The bacteria counts at 14 GMT were always lower than 8GMT and 20GMT counts in the water lettuce, duckweed and algal ponds. There was not much difference in the decline of bacteria in all the 3 pond systems. Enterococci were efficiently removed in all systems (Figures 4-5). It seemed that the different environmental conditions established in all the different systems have been effective in the removal of enterococci.

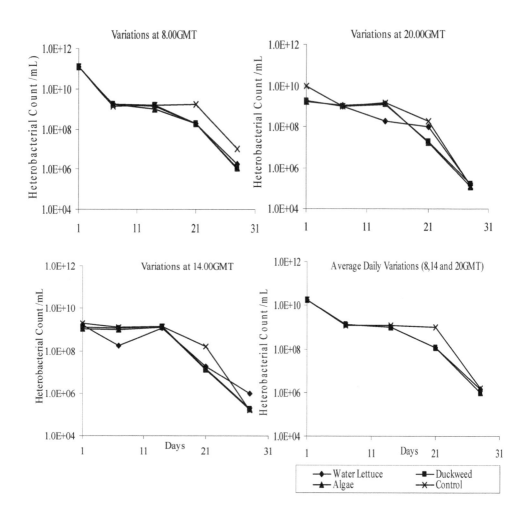

Figure 4 Heterotrophic bacteria in macrophyte and algal ponds

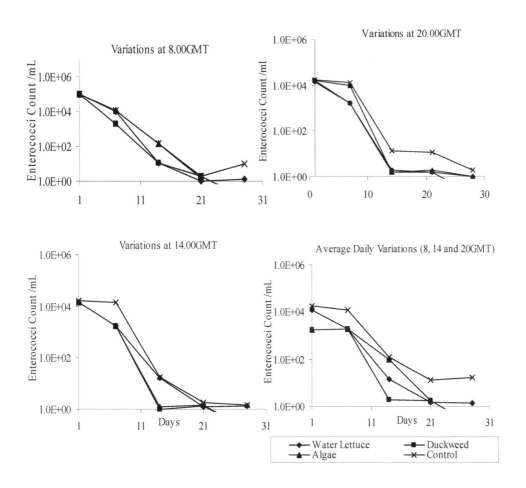

Figure 5 Enterococci bacteria in macrophyte and algal ponds

Organic Matter and Nutrient Removal

All treatment systems removed BOD, COD, ammonia, nitrate and nitrite-nitrogen effectively (Table 1 and Figure 6). Ammonia removal was very high over 99% in all treatment systems. Nitrate removal ranged from 48% in the algal ponds to 62% in the control. Nitrite was completely removed in all treatment systems.

Percentage removal for Organic load removal ranged from 89% in the algal ponds to 96% in the duckweed ponds.

Table 1 Nitrogen removal in macrophyte and algal ponds

Parameter (mg/L)	Initial concentration	Final concentrations after 29 days			
		Control	Water lettuce	Duckweed	Algae
Ammonia-N	51.10	0.40	0.00	0.14	0.00
Nitrate-N	2.21	0.82	0.89	0.96	1.14
Nitrite-N	0.02	0.00	0.00	0.00	0.00

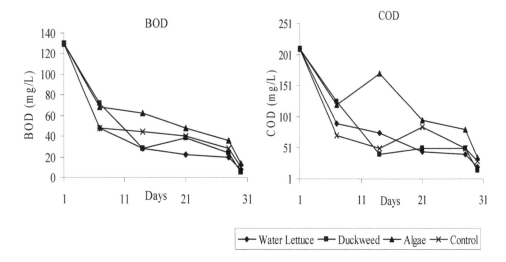

Figure 6 Organic matter in macrophyte and algal ponds

Discussion

Environmental conditions

The alkaline conditions created in the algal ponds observed in this study are well documented in literature (Parhad and Rao, 1974; Curtis *et al.*, 1992; Awuah *et al.*, 2002b). A comparative study of duckweed and algal stabilization ponds in different locations produced alkaline conditions in the algal ponds and neutral conditions in the duckweed ponds (Van der Steen *et al.*, 2000; Zimmo *et al.*, 2002). The environmental conditions in the water lettuce ponds have not been studied in detail. Macauley (1999) observed slightly acidic conditions in water lettuce ponds with pH values of 6.6 in comparison to neutral conditions (7.2) in duckweed ponds. In these comparative studies low oxygen levels were associated with macrophyte ponds, while high oxygen levels were associated with algal ponds. Sridhar and Sharma (1985) also observed acidic conditions in water lettuce ponds in Nigeria. In contrast, Attionu (1976) found alkaline conditions in water lettuce ponds in Ghana. The photosynthetic activities by algae together with respiration created the high pH and DO conditions. The acidic conditions in water lettuce ponds may be due to the release of CO_2 produced during respiration via the roots into the wastewater medium. The gradual increase of pH in the water lettuce and duckweed ponds towards the end of the experiments was due to death of the plants in the absence of nutrients after continuous harvesting. Some spaces were thus exposed, giving room for algae to colonise. In the algal ponds there was no harvesting and most nutrients remained in the pond upon the decay of dead algae to continue the growth of new algae and produce the same environmental conditions.

Heterotrophic and enterococci bacteria removal

The lack of differences in the decline of enterococci in all different systems could be due to different factors operating in each type of system. The die-off in water lettuce ponds could be due to the low pH observed. Awuah *et al*., (2002b) found that acidic conditions were detrimental to enterococci. Garcia and Bécares, (1997) reported higher pathogen removal efficiencies in macrophyte ponds than algal ponds. The macrophytes used were *Typha latifolia, Iris pseudacorus, Scirpus lacustris* and *Phragmites australis*, all of which have very extensive root systems, which they claimed provided a surface area for attachment of the pathogens. Attachment to water lettuce could be a contributing factor in the water lettuce ponds. Spira *et al*., (1981) and Awuah *et al*., (2005) found faecal bacteria attached to floating aquatic macrophytes. Algal ponds are known to be very efficient in the removal of pathogens (Von Sperling, 1996). Feacham *et al*., (1983) reported 100% removal for several pathogens including *Vibrio cholerae* and *Salmonella* in waste stabilization ponds. The die-off in algal pond systems was due to sunlight and strong pH and DO fluctuations observed in the systems (Davies-Colley *et al*., 1999; Awuah *et al*., *unpublished*). Lack of nutrients may have also contributed to die-off of the pathogens (Portier and Palmer, 1989). This is because the batch system used in this study had no refreshments of nutrients. Sunlight is detrimental to bacteria (Pahard and Rao, 1974; Curtis, 1990; Curtis *et al*., 1992). This could have resulted in the lower bacteria counts observed in the afternoons (Figure 5-6). Pearson *et al*., (1987) and Konig (1984) also observed lower bacteria counts in the afternoons and reported also that heterotrophic bacteria numbers actually increased at nighttime. The fluctuations in bacteria can be attributed to photo repair (Liltved and Landfold, 2000), which occurs in the dark after sunlight inactivation. In spite of the possible mechanisms outlined, the enterococci removal in the raw sewage control in the dark did not show any significant difference from the others (Figure 6). The presence heterotrophic bacteria (Figure 5) might have enhanced competition (Wu and Klein, 1976). Long retention times might have also contributed to the natural die-off of enterococci (Figure 5 and 6).

Organic load and nitrogen removal

A high efficiency of organic and nitrogen load removal was observed in all ponds. The removal of organic load in the macrophyte ponds could be due to the presence of oxygen and surface area availability for heterotrophic bacteria to attach and degrade organic materials. In spite of the fact that the algae were not harvested, removal efficiency ranged from 75 to 90% in agreement with Von Spelling's (1996) observation. The reduction was due to sedimentation in both the algae and the raw wastewater (control) to reduce organic load levels. Ammonia-nitrogen and nitrates removal by water lettuce and duckweed were due to plant uptake and during harvesting (Reddy and Debusk, 1987). Nitrification leading to denitrification might have contributed significantly to overall N removal. The high removal efficiencies of BOD and NH_3 which were measured from the supernatant in water lettuce ponds are in line with BOD and NH_3 removal of 83%, 93% respectively, reported by Sharma and Sridhar (1989) in their water lettuce ponds. In addition uptake of ammonia, nitrate and nitrite by algae contribute to nitrogen removal (Lai-P and Lam, 1997). High pH in algal ponds enhances ammonia volatilization (Zimmo *et al*., 2002). The DO levels observed in all systems enhanced the biodegradation of organic matter, which resulted in the low BOD and COD levels observed. The ammonia-nitrogen removal also correlated with organic load removal.

This study shows that in the presence of heterotrophic bacteria and a small amount of oxygen, if the treatment process is given adequate retention time, wastewater will be treated to reduce nitrogen and organic loads and eliminate pathogens.

Conclusion

There were differences in environmental conditions in all treatment systems. Acidic conditions were associated with water lettuce, neutral conditions in duckweed and alkaline conditions in algal treatment systems. Low dissolved oxygen was associated with macrophyte systems and high dissolved oxygen levels above saturation observed in the algal treatment system.

There were no significant differences in the treatment efficiencies of the domestic wastewater treatment systems, with and without macrophytes under these batch experiments. Enterococci removal in all systems was similar even under dark conditions. This shows that mechanisms other than sunlight induced conditions such as pH and dissolved oxygen in pond systems may be responsible for pathogen removal. Competition for nutrients by heterotrophic bacteria, attachment and settling of detached enterococci might have been the main reason behind the removal of pathogens in the treatment systems studied. The interpretation of these results must be done with caution since the indicator organisms used in this study are enterococci and does not represent all groups of pathogens. Besides, the experiment was done on a batch-scale while pond systems are continuous flow.

Acknowledgement

This research was made possible by grants from the Netherlands government through the SAIL foundation.

References

Almasi AU and Pescod MB (1996). Pathogen removal mechanisms in anoxic environment stabilization ponds. *Wat. Sci. Tech.* **33** (7), 133-140.

Attionu RH (1976). Some effects of water lettuce (*Pistia stratiotes* L.) *Hydrobiologia* **50**(3), 245-256.

Awuah E, Nkrumah E and Monney JG (1996). Performance of Asokwa waste stabilization ponds and the condition of other treatment plants in the Ghana. *J. of University of Sci. and Tech.* **16** (1/2), 121-126.

Awuah E, Kuffour AR, Lubberding HJ and Gijzen HJ (2002a). Characterization and management of domestic wastewater in two suburbs of Kumasi. In: *Proceedings of Water and Health Conference*, Ottawa, Canada, September 2002. 475p.

Awuah E, Lubberding HJ, Asante K and Gijzen HJ (2002b). The effect of pH on enterococci removal in *Pistia*, duckweed and algal-based stabilization ponds for domestic wastewater treatment. *Wat. Sci. Tech.* **45** (1), 67-74.

Brix H and Schierup HH (1989). Use of aquatic macrophytes in water pollution control. *AMBIO*, **18** (2), 100-107.

Curtis TP (1990). The mechanisms of pathogen removal of faecal coliforms from waste stabilization ponds. PhD. thesis, University of Leeds, 208p.

Curtis TP, Mara DD and Silva SA (1992). Influence of pH, oxygen, and humic substances on ability of sunlight to damage faecal coliforms in waste stabilization ponds. *Appl. Environ. Microbial.* **58**, 1335-1343.

Curtis TP, Mara DD, Dixo NGH and Silva SA (1994). Light penetration in waste stabilization ponds. *Wat. Res.* **28**, 1031-1038.

Davies-Colley RJ, Donnison AM and Speed DJ (1996). Sunlight wavelengths inactivating faecal indicator microorganisms in waste stabilization ponds. *Wat. Sci. Tech.* **35** (11/12), 219-225.

Davies-Colley RJ, Donnison AM, Speed DJ, Ross CM and Nagels JW (1999). *Wat. Res.* **33**, 1220-1230.

Denny P (1997). Implementation of constructed wetlands in developing countries. *Wat. Sci. Tech.* **35**(5), 22-34

Feachem RG, Bradley DJ, Garelick H and Mara DD (1983). Sanitation and Disease: Health Aspects of Excreta and wastewater management. Published for the World Bank by John Wiley and Sons, U.K.

Garcia M and Bécarès E (1997). Bacterial removal in three pilot scale wastewater treatment systems for rural areas. *Wat. Sci. Tech.* **35** (11/12), 197-200.

Gijzen HJ and Ikramullah M, (1999). Pre-feasibility of duckweed-based wastewater treatment and resource recovery in Bangladesh. Final Report. World Bank, pp. 87.

Gijzen HJ and Khonker M (1997). An overview of the ecology, physiology, cultivation and applications of duckweed. Inception Report. Annex literature review. Duckweed Research Project (WRP). Dhaka, Bangladesh. World Bank 53p.

Greenberg AE, Clesceri LS and Eaton AD (1992). Standard methods for the examination of water and wastewater. 18th Edition. Water Environmental Fed. Washington D.C.

Iqbal S (1999). Duckweed aquaculture potentials, possibilities and limitations for combined wastewater treatment and animal feed production in developing countries. SANDEC Report No.6/99, pp. 91.

Konig A (1984). Ecophysiological studies on some algae and bacteria of waste stabilization ponds PhD Thesis University of Liverpool.

Lai-P CC and Lam PKS (1997). Major pathways for nitrogen removal in wastewater stabilization ponds. *Wat. Air, Soil Pollut.* **94**(1/2), 125-136.

Liltved H and Landfold B (2000). Effects of high intensity light on ultraviolet–irradiated and non-irradiated fish pathogenic bacteria. *Wat. Res.* **34**, 481-486.

Macauley M (1999). The role of attachment in the removal of pathogens from macrophyte-based wastewater stabilization ponds. MSc Thesis D.E.W. 092 UNESCO-IHE. Delft, The Netherlands.

Mandi L, Ouzzani N, Bouhoum K and Boussaid A (1993). Wastewater treatment by stabilization ponds with and without macrophytes under arid climate *Wat. Sci. Tech.* **28** (10), 177-181.

Ministry of Health, (1996). Health Sector Facts and figures, pp. 31.

Nhapi I, Dalu J, Ndamba J, Siebel MA and Gijzen HJ (2003). An evaluation of duckweed-based pond systems as an alternative option for decentralised treatment and re-use. *Wat, Sci. Tech.* **48**(2), 323-330.

Niemi RM and Ahtiainen J (1995). Enumeration of intestinal enterococci and interfering organisms with Slanetz-Bartley agar, KF streptococcus agar and the MUST method. *Lett. Appl. Microbiol.* **20**(2), 92-7.

Oron G, Wildschut LR and Porath D (1985). Wastewater recycling by duckweed for protein production and effluent renovation. *Wat. Sci. Tech.* **17** (4/5), 803-817.

Parhad NM and Rao NU. (1972). The effect of algal growth on survival of *E. coli* in sewage. *Indian J. Environ. Hlth.* **14** (2), 131-139.

Parhad NM and Rao NU (1974). Effect of pH on the survival of *Escherichia coli.* *Wat. Pollut. Contr. Fed* **55** (1), 285-296.

Pearson HW, Mara DD, Mills SW and Smallman DJ (1987). Physicochemical parameters influencing faecal bacteria survival in waste stabilization ponds. *Wat. Sci. Tech.* **19**(12), 145-152.

Portier R and Palmer S (1989). Wetlands microbiology: Form, function, process. In: Hammer DA (ed). *Constructed wetlands for wastewater treatment: Municipal, industrial and agricultural*. Lewis publishers Chelsea, Michigan. pp. 89-106.

Reddy KR and Debusk TA (1987). State of utilization of aquatic plants in water pollution control. *Wat. Sci. Tech.* **19** (10), 61-79.

Salifu LY (1996). Appraisal of sewage maintenance in Ghana. In: Proceedings of 2[nd]West Africa Water, Sanitation and Environment conference,6p.

Sharma BM and Sridhar MKC (1985). Some observations on the oxygen changes in a lake covered with *Pistia stratiotes*. L. *Wat Res.* **19**, 953-939.

Sharma BM and Sridhar MKC (1989). Growth characteristics of water lettuce (*Pistia stratiotes* L) in Southwest Nigeria. *Archiv fuer hydrobiologie AHYBA4:* **115** (2), 305-312.

Spira WM, Hug A, Ahmed QS and Saeed YA (1981). Uptake of *Vibrio cholerae* Biotype *eltor* from contaminated water by water hyacinth (*Eichhornia crassipies*). *Appl. Env. Microbiol.* **42,** 50-553.

UNICEF/UNEP (1990). The state of the art of the environment-1990. 73p.

Van der Steen P, Brenner A, Shabtai Y and Oron G (2000). Improved faecal coliform decay in integrated duckweed and algal ponds. *Wat. Sci. Tech.* **42** (10/11), 363-370.

Von Sperling M (1996). Comparison among most frequently used systems for wastewater treatment in developing countries. *Wat. Sci. Tech.* **33** (3), 59-76.

Wu S and Klein DA (1976). Starvation effects on *Escherichia coli* and aquatic bacteria to nutrient addition and secondary warming stresses. *Appl. Environ. Microbiol.* **31**, 216-220.

Zimmo OR, van der Steen P, Al-Saéd RM and Gijzen HJ (2002). Process performance assessment of algae-based and duckweed wastewater treatment systems. *Wat. Sci. Tech.* **45**(1), 91-102.

Chapter Three

The Effect of pH on Enterococci Removal in Water Lettuce, Duckweed and Algal Ponds

Adapted from
Awuah E, Lubberding HJ, Asante K, and Gijzen HJ (2001)
The effect of pH on enterococci removal in Pistia duckweed and algae-based stabilization ponds
for domestic wastewater treatment
Water Science Technology **45**(1), 67-74.

The Effect of pH on Enterococci Removal in Water Lettuce, Duckweed and Algal Ponds

Abstract

A batch scale experiment was conducted to determine the effect of pH on enterococci removal rate. The batch experiment was conducted using pH 4, 5, 7, 9, and 11 treatments under light and dark conditions. Enterococci concentrations and DO were measured every day and every other day respectively for nine days. A bench scale continuous system was also constructed to determine the environmental conditions and its effects on enterococci removal. The system comprised of water lettuce, duckweed and algal ponds operating in parallel with each system having a series of four ponds and a total retention period of 28days after two days of anaerobic pre-treatment. After two months of operation, temperature, pH, DO, TDS, and enterococci populations were monitored. Enterococci die-off rates at pH 4, 5, 7, 9 and 11 in the light/dark were (expressed as kd^{-1}); 2.1/2.1, 2.1/1.5, 2.1/1.5, 2.1/1.4 and 1.1/1.0, respectively. DO levels in the batch incubations were low, between 0.17 mg/L at pH 4 (light) to 0.56mg/L at pH 7 (light). Low pH of 4.4 was obtained in the water lettuce pond system. Neutral conditions were observed in duckweed pond system. pH values > 9 were observed in the algal pond system. Enterococci decreased from 7.8×10^6 to < 500/100mL in all pond systems after 28 days of treatment with no significant differences between treatments.

Keywords: Water lettuce; duckweed; algae; enterococci; pH

Introduction

pH is an important environmental parameter in wastewater treatment. Aquatic organisms, which play a role in treatment processes, are sensitive to pH changes, and biological treatment requires either pH control or monitoring. Elevated pH has been found to contribute significantly to faecal coliform removal in waste stabilization ponds (Pearson *et al.*, 1987). Curtis *et al.*, (1992) have similarly reported that high and fluctuating levels of pH and dissolved oxygen (DO) in algal ponds are detrimental to pathogens in wastewater. Earlier studies by Awuah *et al.*, (2001) on the environmental conditions and enterococci removal in macrophyte and algal ponds on a batch scale showed that there were no differences in the enterococci removal in all pond systems even though the environmental conditions were different. A low pH of 4.4 was observed in the water lettuce ponds, while neutral conditions were observed in the duckweed ponds. In the algal ponds, a high pH of 10.5 was observed. The control (raw sewage kept in the dark) also had a comparatively high level of pathogen removal. Most research work on the effect of pH on pathogens in wastewater has focused mainly on high pH and on faecal coliforms. The use of enterococci as an indicator organism has not been widely tested.

Our research is focused on determining the pathogen removal mechanisms in macrophyte and algal ponds. The objectives of this study are to determine the effects of acidic, neutral and basic ranges of pH on pathogen removal rate using enterococci as indicator organisms and to determine the environmental conditions and enterococci removal of macrophyte and algal ponds in a continuous flow system.

Materials and Methods

The experiments were carried out at the Kwame Nkrumah University of Science and Technology (KNUST), Kumasi, Ghana. Effect of pH on enterococci removal was conducted on a batch scale in 1L-opaque white plastic containers. A grab sewage sample obtained from the University treatment plant was allowed to undergo anaerobic treatment for two days. Eight hundred milliliters of this anaerobically pre-treated sewage was put into each of thirty 1-L plastic containers. Two sets of five treatments consisting of pH 4, 5, 7, 9 and 11 were made in triplicates. One set of the five treatments was arranged in a randomized block design and placed in moistened sand boxes (to regulate temperature in the containers) and exposed to sunlight. The other set was kept in a dark cupboard in the laboratory to mimic conditions in macrophyte ponds. The pH in the pond systems was adjusted daily with 0.2N HCl and 0.2N NaOH solutions. Enterococci populations were measured daily while DO were monitored every other day for 9 days.

A bench-scale continuous flow system was constructed to simulate continuous flow waste stabilization ponds. The set up was made up of an anaerobic pond, followed by three parallel lines of four ponds consisting of water lettuce (*Pistia stratiotes*) duckweed (*Spirodela polyhriza*) and algae (natural colonization). Ponds were made of large opaque plastic containers with a depth of 0.63m. Each pond had a retention period of 7 days and each pond system had a total retention period of 28 days excluding 2 days retention period in the anaerobic pond. A flow rate of $0.01m^3$ a day was maintained in all pond systems by gravity.

After 2 month start-up period, environmental parameters and enterococci numbers were monitored. Temperature, pH, DO and total dissolved solids (TDS) were monitored every other day at 8, 13 and 18 Greenwich Meridian Time (GMT) and at depth of 0.10m (surface), 0.35m (middle) and 0.63m (bottom layer).

Enterococci populations of the effluents in the bench-scale continuous flow set up from each pond were monitored five times during the study period. Temperature and pH measurements were taken with a portable electronic pH meter. DO levels in the various treatments were monitored with portable electronic probe microprocessor oximeter. Total dissolved solids were measured with a Microprocessor conductivity meter.

The enterococci populations were determined using the spread plate method on a Slanetz and Bartley medium at 44.5°C (Niemi and Ahtiainen, 1995). The enterococci removal rate constant was calculated by using Chick's law. The MSTAT statistical package was used to analyze the results.

Results

Batch-scale

The DO levels in the batch scale pH experiment were generally low. The treatments exposed to sunlight generally had higher values than those kept in the dark (Table1). In light, pH 7 had the highest DO of 0.56mg/L while the lowest DO of 0.17mg/L was recorded at pH 4 in light (Table 1). There were no significant differences between the mean DO readings recorded in the dark.

Table 1 Daily mean DO and enterococci removal rates in light and dark conditions

pH	Mean DO (mg/L)		Removal rates (kd^{-1})	
	Light	Dark	Light	Dark
4	0.17a*	0.21a	2.1a	2.1a
5	0.40b	0.30d	2.1a	1.5b
7	0.56c	0.22a	2.1a	1.5b
9	0.33d	0.22a	2.0a	1.4b
11	0.43b	0.25a	1.1c	1.0c

Different letters means significant differences (p<0.05) in each table

No enterococci were detected after the fifth day and beyond at pH 4, 5 and 7 in the set exposed to light. The incubation at pH 9 showed the presence of enterococci on day six but not on the fifth or after day 6. All incubations under dark condition showed the presence of enterococci till the 9th day (Figure 1). pH 4 treatment had high die-off rate constants of 2.1 k^{-d} under both light and dark conditions.

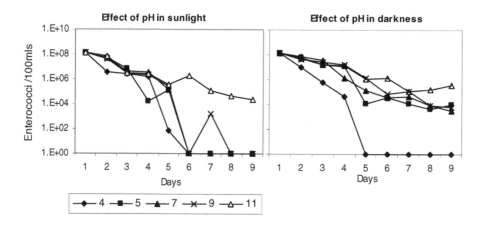

Figure 1 Enterococci removal at different pHs in domestic wastewater

Continuous flow system

Generally pH was acidic, neutral and alkaline in the water lettuce, duckweed and algal ponds, respectively. pH did not vary significantly (P<0.05) with depth (Table 2).

Diurnal pH variations within individual ponds were also not significant (P<0.05) even though higher pH was observed in the algal ponds, especially at 13GMT than at 8GMT and 18GMT (Table 3). The pH of water lettuce ponds decreased from about 6.4 in the first pond to 4.4 after 28 days retention period (Table 2 and 3). In the algal ponds, pH increased from 7.5 in the first pond to > 9 in the 4th pond. Highest mean pH of 9.5 was observed in this pond around mid-day (Table 3). Higher DO values were recorded near the surface than in the sediments. DO concentrations also increased during the later stages of treatment (ponds 3 and 4), reaching peaks of 3.1, 3.8 and 10.5mg/L for duckweed, water lettuce and algal ponds, respectively (Table 2 and 3). Diurnal fluctuations were greatest in the algal ponds.

Evaluation of the effluent from each of the four ponds in series of each pond system showed that the longer the hydraulic retention time, the higher the efficiency of removal. The enterococci concentration declined with time in all the pond systems from an initial concentration of 7.8×10^6/100 m/L in the anaerobic pond to values below 500/100m/L in the last pond (Figure 2).

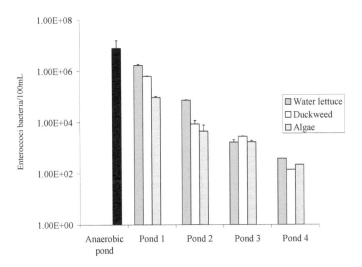

Figure 2 Removal of enterococci in macrophyte and algal ponds

Table 2 pH, DO and TDS at different depths in pond systems

Parameter	Treatment	Pond number	0.10m	0.35m	0.63m
pH	Water lettuce	1	6.3±0.0	6.4+±0.1	6.4±0.1
		2	5.6±0.0	5.7±0.0	5.7±0.0
		3	4.9±0.0	4.9±0.0	4.9±0.0
		4	4.4±0.2	4.6±0.1	4.4±0.2
pH	Duckweed	1	7.0±0.0	7.0±0.0	7.0±0.0
		2	6.9±0.5	7.0±0.0	7.0±0.1
		3	6.8±0.0	6.9 ±0.0	6.9±0.0
		4	6.7±0.0	6.7±0.1	6.7±0.0
pH	Algae	1	7.8±0.2	7.6±0.2	7.4±0.2
		2	7.8±0.2	7.4±0.2	7.4±0.2
		3	7.6±0.3	7.4±0.3	7.0±0.3
		4	9.1±0.4	8.9±0.4	8.4±0.4
DO (mg/L)	Water lettuce	1	1.0±3.4	0.7±0.8	0.5±0.2
		2	1.0±0.6	0.8±1.7	0.7±0.2
		3	2.6±0.6	1.9±2.1	1.7±0.2
		4	3.7±0.7	3.6±2.5	2.3±0.2
DO (mg/L)	Duckweed	1	1.0±3.5	0.9±0.1	0.7±0.2
		2	1.5±3.2	1.2±0.6	0.7±0.2
		3	2.2±0.8	1.9±2.1	1.0±0.2
		4	2.9±0.3	2.8±0.1	2.2±0.2
DO (mg/L)	Algae	1	6.2±0.1	1.7±0.1	0.8±0.2
		2	5.9±0.4	1.6±0.4	0.8±0.1
		3	5.5±1.5	1.4±1.6	0.9±2.6
		4	10.4±2.7	5.3±0.1	2.1±2.4
TDS (mg/L)	Water lettuce	1	255±14	257±10	256±5
		2	132±3	34±5	134±4
		3	96±2	96±2	96±2
		4	90±3	91±2	91±2
TDS (mg/L)	Duckweed	1	548+21	564±3	556±8
		2	444+26	444±3	451±8
		3	352±2	352±4	356±3
		4	260±3	261±2	262±2
TDS (mg/L)	Algae	1	406±20	410±14	421±10
		2	341±41	348±11	361±11
		3	286±8	304±11	380±10
		4	290±22	274±7	278±7

* Average values of readings at different times (08GMT, 13GMT and 20GMT)

Table 3 Diurnal variations of environmental conditions in macrophyte and algal ponds*

Parameter	Treatment	Pond number	8GMT Morning	13GMT Afternoon	18GMT Evening
pH	Water lettuce	1	6.3±0.0	6.4±0.0	6.3±0.0
		2	5.7±0.0	5.7±0.0	5.6±0.0
		3	4.8±0.1	4.8±0.1	5.0 ±0.1
		4	4.4±0.2	4.3±0.2	4.5±0.1
pH	Duckweed	1	6.9±0.0	7.0±0.0	7.1±0.0
		2	7.0±0.0	6.9±0.0	7.0±0.0
		3	6.8±0.0	6.8±0.0	7.0±0.0
		4	6.7±0.0	6.7±0.0	6.8±0.0
pH	Algae	1	7.4±0.1	8.0±0.1	7.5±0.0
		2	7.3±0.1	7.8±0.1	7.4±0.1
		3	7.1±0.4	7.8±10.3	7.1±0.3
		4	7.9±0.3	9.5±0.3	8.9±0.3
DO (mg/L)	Water lettuce	1	0.6±0.1	0.6±0.1	1.0±0.2
		2	0.7±0.1	0.7±0.1	1.1±0.6
		3	1.3±0.1	2.6±0.1	2.3±0.4
		4	2.4±0.3	3.9±0.3	3.4±0.3
DO (mg/L)	Duckweed	1	0.5±0.2	0.7±0.4	1.4±2.1
		2	0.7±1.6	1.5±0.2	1.1±2.1
		3	1.1±0.2	2.0±1.9	1.9±3.1
		4	2.0±0.1	3.1±0.1	2.8±2.0
DO (mg/L)	Algae	1	0.7±0.2	6.3±0.2	1.6±0.5
		2	0.9±0.2	5.8±0.3	1.6±0.9
		3	0.8±3.1	5.8±0.2	1.3±2.5
		4	2.4±2.3	5.8±0.3	6.9±3.4
TDS (mg/L)	Water lettuce	1	259±4	266±2	245±3
		2	134±4	136±2	130±4
		3	95±3	95±3	94±2
		4	91±2	91±1	90±2
TDS (mg/L)	Duckweed	1	534±6	554±2	566±3
		2	426±5	454±3	459±2
		3	341±5	352±2	368±3
		4	258±3	261±3	266±4
TDS (mg/L)	Algae	1	388±10	425±5	432±6
		2	325±12	372±9	353±4
		3	288±9	319±10	293±5
		4	288±6	275±6	279±8

* *Average values at all depths (0.1m, 0.35m and 0.63m)*

Even though alkaline conditions were observed in algal ponds, it performed better ($P<0.05$) in enterococci removal than the water lettuce and duckweed systems in the first and second ponds (Figure. 2). A single factor analysis of variance, however, showed that there were no significant ($P<0.05$) differences in the three pond systems after 28 days. There were however, significant differences between water lettuce ponds and duckweed and algal ponds after 7 and 14 days in the removal of enterococci. No significant differences in the removal of enterococci were observed between water lettuce ponds and that of duckweed and algal ponds after 21 days. There were also no significant differences between duckweed and algal ponds after 14 days and 21 days.

Discussion

Batch scale

High pH is known to excite molecular oxygen to the ionic form, which is toxic (Zepp *et al* 1981; Haag *et al* 1986). However, DO levels recorded in this study were too low for this to play a meaningful role in enterococci removal. Van Buuren and Hobma (1991) observed the limit of 0.5mg/L to be bactericidal. The highest DO observed in the incubation at pH 7 (0.56mg/L) was very low in comparison to 8-15 mg/L reported by Baumgartner (1996).

The better survival of enterococci than faecal coliform at high pH could be due to fact that the former is gram-positive and the later are gram-negative. The cell walls are accordingly different. The thick murein structure and other unknown factors within the cell wall of enterococci may protect the cell and cause resistance to high pH enabling survival at pH values higher than 9. Brock, (1979) reports on the ability of enterococci to grow at pH 9.6.

The values of these removal rates obtained are comparable to values obtained in our previous experiments using water lettuce, duckweed and algae at 2.4 d^{-1}, 2.1 d^{-1}, 2.2 d^{-1} respectively (Awuah *et al*, 2001). In the control, (raw sewage exposed to dark conditions) the removal rate was 2.1 d^{-1}. The values in this study are higher than removal rates reported in literature for faecal coliforms in maturation ponds and the recommended value for design of treatment plants. Saqqar and Pescod (1992) found the removal rate of faecal coliforms in maturation ponds to be 0.33 d^{-1}at 14^0 C and 0.93 d^{-1}at 24^0C. According to Johansson *et al.,* (1996), the recommended removal rate for faecal coliforms in the design of treatment plants is -1.06 d^{-1}. The temperature range under which this present study was conducted was between 27 and 37 ^{0}C. The high pH tolerance exhibited by enterococci may also explain why Davies-Colley *et al.,* (1994) found low removal of enterococci in seawater which has high pH.

Light might have played a role in enterococci die-off (Table1). According to Curtis (1990) and Curtis *et al.,* (1992) sunlight causes damage to the cytoplasmic membranes of bacteria, which makes the organism more sensitive to the effects of other factors such as pH. Many researchers including Gersberg *et al.,* (1987), Frijns and Nooteboom (1989) and Davies-Colley *et al.,* (1997) have shown that sunlight is lethal to faecal coliforms and enterococci and enhances their die-off rate. At pH 4 however, light did not seem to have any effect on the ability of this pH to destroy enterococci.

Other researchers have also shown that high pH (>9) was detrimental to bacteria survival (Parhad and Rao, 1974; Pearson *et al.,* 1987). According to Hirn *et al.,* (1980), pH >9 was the main or even the sole factor affecting the removal rate of faecal coliforms. Troussellier *et al.,* (1986) using critical path analysis also showed that light and high pH were the key causal effects controlling bacterial removal in sewage lagoons. Pearson *et al.,* (1987) reported that faecal coliforms appeared to be adversely affected by high dissolved oxygen and high pH. Zepp *et al.,* (1981) and Haag *et al.,* (1986) explained that high pH decreases the resistance of the bacteria to the bactericidal effect of light. According to these authors, high pH may increase the production of toxic forms of oxygen, which damages the cytoplasmic membrane, allowing hydroxyl ions to enter. This would cause an increase in the internal pH of the bacteria and kill them ultimately. This did not seem to have taken place in the case of enterococci removal in the presence o sunlight at high pH.

Continuous flow systems

The environmental conditions created in all the systems seem to be sufficiently detrimental for enterococci removal and no single factor seems to dominate removal rate of enterococci. The removal rate may be due to a combination of factors. These may include the long retention period (Oragui *et al.*, 1987), pH (Parhad and Rao, 1974), high dissolved oxygen levels (Curtis, 1990), presence of predators (Ellis, 1983), algal toxins (Toms *et al.*, 1975), sedimentation (Gannon *et al.*, 1983) and sunlight and UV light (Davies-Colley *et al.*, 1997). Pearson *et al.*, (1987) observed that at high DO levels, enterococci die faster. Aeration enhances faecal coliform removal rates in sewage effluent (Klock, 1971). Sudden changes in oxygen concentration increased the removal rate of *Salmonella typhimurium* strains (Barzily *et al.*, 1991). Since the DO was higher in all the systems, than the minimum lethal level (0.5mg/L) as quoted by Van Buuren and Hobma (1991), there could be a significant effect of the DO on enterococci removal.

The enterococci removal in the water lettuce ponds could be due to the acidic conditions observed in the last three ponds in series. In the duckweed systems, the major contributing factor could be the long retention period. The long retention period applies to all the pond systems. Without the long detention period, none of the three pond systems would have achieved enteroccocci concentrations of < 500/100mL (Figure 2), which was the value obtained at the end of the 28 days for all treatment systems. In the algal ponds, removal was probably due to direct sunlight penetration and fluctuations in DO and pH. Davies-Colley *et al.*, (1997) found enterococci to be highly susceptible to sunlight. The presence of predators and competition for nutrients could also play a role (Wu and Klein, 1976; Legendre *et al.*, 1984). However, measurements of protozoa populations were not done in this study.

Despite the high enterococci removal in the batch scale experiments, the continuous flow system did not remove all the enterococci. This was probably due to short-circuiting. Besides, environmental conditions in the continuous flow system were different from those observed under the batch type incubations. Other authors have also reported lower removal rates of indicator organisms in full-scale stabilization ponds than in batch type incubations (Curds and Fey, 1969; Pearson *et al.*, 1987). The enterococci removal rate in all the pond systems was due to a combination of factors discussed above. The fact that environmental conditions were different in the three systems, yet no significant differences in removal were observed after 28 days of treatment suggests that different mechanisms could be involved in algal and macrophyte ponds for enterococci removal.

Conclusion

Low pH has more bactericidal effects on enterococci in domestic wastewater than alkaline conditions. The bactericidal properties at different pH values are enhanced by sunlight. Dissolved oxygen might not have played any significant role in the pathogen removal rate in the batch scale study. The study also shows that enterococci can survive at a high pH of 11 especially under dark conditions. Differences in environmental conditions exist in macrophyte and algal ponds. Water lettuce ponds produce acidic, duckweed ponds neutral and algal ponds alkaline conditions. The presence of oxygen, light, nutrient depletion, low pH and the long retention period in the continuos flow systems might all have played a role in the removal of enterococci in macrophyte and algal ponds. Pathogen removal mechanisms in macrophyte and algal ponds are different. The effect of pH on faecal coliforms and other pathogen indicators, under similar conditions should be studied for comparison.

Acknowledgement

This research is made possible by grants from the Netherlands government through the Sail foundation.

References

Awuah E, Lubberding HJ, Asante K and Gijzen HJ (2001). Environmental conditions and pathogen removal in macrophyte and algal-based domestic wastewater treatment systems. *Wat. Sci. Tech* **44**(6), 11-18.

Baumgartner DJ (1996). Surface water pollution. In: Pepper JL, Gerba PC and Brusseau ML (eds). *Pollution Science*. Academic Press Inc. Canada, pp. 189 – 220.

Brock TD (1979). Biology of microorganisms. 3rd edition. Prentice-Hall, Inc. New Jersey, 802p.

Curds CR and Fey GJ (1969). The effect of ciliated protozoa on the fate of *Escherichia coli* in the activated sludge process. *Wat. Res.* **3**, 853-867.

Curtis TP (1990). The mechanism of faecal coliform removal from waste stabilization ponds. PhD thesis, University of Leeds, 208p.

Curtis TP, Mara DD and Silva SA (1992). The effect of sunlight on faecal coliforms in ponds: implications for research and design. *Wat. Sci. Tech.* **26** (7/8),1729-1738.

Davies-Colley R, Bell R, and Donnison A (1994). Sunlight inactivation of faecal coliforms and faecal streptococci in sewage effluent attributed to seawater. *Appl. Environ. Microbiol.* **60**, 2049-2058.

Davies-Colley R, Donnison A and Speed, DJ (1997). Sunlight wavelengths inactivating faecal indicator microorganisms in waste stabilization ponds. *Wat. Sci. Tech.* **25** (5), 219-225.

Ellis K (1983). Stabilization ponds: design and operation. *Crit. Re. Env. Contr.*, **13**, 69-102.

Frijns J and Nooteboom L (1989). Afbraak van E-coli in Huishoudelijk afvalwater. MSc thesis. Dept. Environ. Tech., Wageningen Agricultural University, The Netherlands.

Gannon JJ, Buse K, and Schillinger, J (1983). Faecal coliform disappearance in a river impoundment. *Wat. Res.*, **17**,1595-1601.

Gersberg RM, Lyon SR, Brenner R and Elkins BV (1987). Fate of viruses in artificial wetlands. *Appl. Environ. Microbiol.* **83**, 731-736.

Haag WR, Hoigne J, Gassman E and Bruun A (1986). Singlet oxygen in surface waters, part III. Photo-chemical formation and steady state concentrations in various types of water. *Chemosphere* **13**, 641-650.

Hirn J, Viljamaah H and Raevuorsi M (1980). The effect of physicochemical, phytoplankton and seasonal factors on faecal indicator bacteria in northern brackish water. *Wat. Res.* **14**, 279-285.

Johansson P, Penrup M and Rangeby M (1996). Low-cost upgrading of an oversized waste stabilization pond system in Mindelo Cape Verde. *Wat. Sci. Tech.* **33**(7), 99-106.

Kaneko M (1997). Virus removal by the domestic wastewater pond system named Johkasou. *Wat. Sci. Tech.* **35**(5), 187-191.

Klock JW (1971). Survival of faecal coliforms in wastewater. *Wat. Pollut. Contr. Fed.* **50**, 20171-20183.

Legendre P, Baleux B and Troussellier M (1984). Dynamics of pollution indicator and heterotrophic bacteria in sewage treatment lagoons. *Appl. Environ. Microbiol.* **48**, 486-593.

Niemi RM and Ahtiainen J (1995). Enumeration of intestinal enterococci and interfering organisms with Slanetz-Bartley agar, KF streptococcus agar and the MUST method. *Lett. Appl. Microbiol.* **20**(2), 92-7.

Oragui JI, Curtis T, Silva SA and Mara DD (1987). The removal of faecal coliforms in experimental waste stabilization pond systems with different geometries and configuration. *Wat. Sci. Tech.* **19** (3/4), 569-573

Parhad NM and Rao NU (1974). Effect of pH on the survival of *E. coli*. *Wat. Pollut. Contr. Fed.* **55**, 285-296.

Pearson HW, Mara DD, Mills, SW and Smallman, DT (1987). Physico-chemical parameters influencing faecal bacterial survival in waste stabilization ponds. *Wat. Sci. Tech.* **19**(5), 145-152.

Portier R and Palmer S (1989). Wetlands microbiology: Form, function, process. In: Hammer DA (ed) *Constructed wetlands for wastewater treatment: Municipal, industrial and agricultural*, Lewis Pub. Chelsea, Michigan. pp. 89-106.

Toms JP, Owens M, Hall JA and Mindenhall, MJ (1975). Observations on the performance of polishing lagoons at a large works. *Wat. Pollut. Contr.* **74**, 383-401.

Troussellier M, Legendre P and Baleux B (1986). Modeling of the evolution of bacteria densities in an eutrophic ecosystem (sewage lagoons). *Microbial Ecol.* **12**, 355-379.

Saqqar MM and Pescod MB (1992). Modeling coliform reduction in wastewater stabilization ponds. *Wat. Sci. Tech.* **26** (7/8), 1667-1677.

Van Buuren JCL and Hobma S (1991). The faecal coliform removal rate at post treatment of anaerobically pre-treated domestic wastewater. Department of Environ. Tech. Agricultural University, Wageningen. *Unpublished*.

Wu S and Klein DA (1976). Starvation effects on *Escherichia coli* and aquatic bacterial to nutrient addition and secondary warming stresses. *Appl. Environ. Microbiol.* **31**, 216-220.

Zepp RG, Baughman GL and Schlotzhauer PF (1981). Comparison of photo-chemical behaviour of various humic substances in water. II Photosynthesized Oxygenations. *Chemosphere* **10**, 119-126.

Chapter Four

Environmental Conditions and Effect of pH on Faecal Coliform Removal in Macrophyte and Algal Ponds

Awuah E, Boateng J, Lubberding HJ and Gijzen HJ
Environmental Conditions and Effect of pH on Faecal Coliform Removal in
Macrophyte and Algal Ponds

Environmental Conditions and Effect of pH on Faecal Coliform Removal in Macrophyte and Algal Ponds

Abstract

The environmental conditions within 3 pond systems comprising of water lettuce (*Pistia stratiotes*), duckweed (*Spirodela polyrhiza*) and algae (natural colonization), each operating in a series of four ponds were assessed specifically for temperature, pH, DO and total dissolved solids. The pH readings observed in the pond systems were simulated to determine the effect of different pH (4, 5, 7, 9 and 10), on faecal coliform removal in domestic wastewater. Acidic conditions prevailed in the water lettuce ponds and ranged from 4.3 to 6.3, duckweed ponds were neutral with values between 6.7 and 7.1. The algal ponds were mostly alkaline with pH values ranging from 7 to 10. The DO levels were low in the macrophyte ponds. The algal ponds had saturated levels of DO. The removal of faecal coliforms was lowest in the water lettuce ponds followed by duckweed and algal ponds respectively. The removal rates of faecal coliform at pH 7, 9 10 and the raw wastewater without pH adjustment under both sunlight and dark conditions were not significantly different. Removal rates at pH 5 were the lowest among all treatments. Under light conditions k (d^{-1}) values of 1.5, 0.2, 1.3, 1.4, 1.4 and 1.4 were observed respectively for pHs of 4, 5, 7, 9, and 10 treatments and for the raw wastewater without pH adjustments. Under dark conditions the removal rates were 1.1, 0.5, 1.0, 1.2, 1.4 and 1.2 respectively for pH 4, 5, 7, 9, and 10 treatments and the raw wastewater pH without adjustment.

Key words Water lettuce; duckweed; algae; faecal coliforms; pH

Introduction

Pathogen removal in domestic wastewater is essential to curtail the incidence of enteric diseases. Numerous physical and chemical parameters have been suggested as factors involved in pathogen removal in wastewater treatment systems. These include temperature, light, dissolved oxygen, pH, nutrient depletion, predation, attachment and coliphages attack (Maynard *et al.*, 1999).

Waste stabilization ponds (WSPs) have been widely studied and their efficiencies in providing effluent of good quality are acknowledged. Additionally, WSPs are cheap in operation and can achieve effective removal of pathogens, without the negative side effects of chlorination (Feacham *et al.*, 1983; Von Sperling, 1996). However, there is still much debate on the mechanisms of bacterial removal, and there have been conflicting reports on the contributions of the different physical and chemical factors involved (Davies-Colley *et al.*, 1999; Maynard *et al.*, 1999). In a previous research by Awuah *et al.*, (2001; 2002) the removal of enterococci as indicator organism in continuous flow and batch scale studies of macrophyte and algal ponds was attributed to natural die-off, alkaline and acidic pH and dissolved oxygen in the treatment systems. Sunlight was found to be more detrimental to enterococci than darkness. Since faecal coliforms are more widely used as indicator organisms than enterococci, it is necessary to know the response of faecal coliforms to similar environmental conditions.

The aim of this study therefore, was to assess the effect of prevailing environmental conditions within macrophyte and algal-based ponds and determine the effect of pH on the removal of faecal coliforms

Materials and Methods

Environmental conditions and pond performance in faecal coliform removal

The research was carried out at the Kwame Nkrumah University of Science and Technology (KNUST), Kumasi, Ghana.

Waste stabilization pond systems treating domestic wastewater of KNUST comprising of water lettuce (*Pistia stratiotes*), duckweed (*Spirodela polyrhiza*) and algal (natural colonization) treatment ponds operating in series (Figure 1) were studied extensively for eight weeks. The system consisted of 3 parallel treatment lines of 4 ponds each, involving the use of water lettuce (*Pistia stratiotes*), duckweed (*Spirodela polyrhiza*) and algae (natural colonization). A flow rate of $0.01m^3$/day was maintained in each treatment system. Each pond had a depth of 0.63m and surface area of $0.145m^2$ and hydraulic retention period of 7 days. Wastewater was collected at the influent grit chamber of the university's wastewater treatment plant at 7 GMT daily and put into the anaerobic pond, which fed into the 3 pond systems, by gravity in a continuous flow (Figure 1). The water lettuce ponds were maintained by harvesting macrophytes once a very week. Duckweed ponds were harvested twice in a week. The performance in terms of faecal coliform removal and the prevailing environmental conditions were assessed. The following environmental conditions were assessed: Temperature, pH, dissolved oxygen (DO), and total dissolved solids (TDS). Measurements were made at 8, 13 and 20 GMT at the surface (0.1m), middle (0.35cm) and bottom (0.63m) depths of all ponds. Faecal coliform measurements were made on the effluents of each pond in each stage of treatment in the 3 pond systems Sampling was done in the morning before 9 GMT.

Temperature, pH, conductivity, TDS and DO were measured with portable electronic probe meters (LF 323 — B/ SET 2, WTW – Germany). Faecal coliforms were determined using membrane lauryl sulphate broth and confirmed in EC medium (Greenberg *et al.*, 1992).

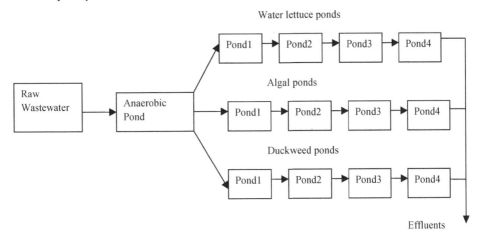

Figure 1 Schematic diagram of bench-scale macrophyte and algal waste stabilization ponds

Effect of pH on faecal coliform removal

To mimic the continuous flow system with respect to pH, the effect of pH ranges observed in the different pond systems on faecal coliform removal using domestic wastewater that had undergone anaerobic digestion for 2 days was studied. 800mL of pre-digested wastewater was poured into each of 36, 1-litre plastic containers, Two sets of 6 treatments (pH values: 4, 5, 7, 9, 10 and raw wastewater without pH adjustment) in three replicates were arranged randomly using the randomised block design in Table 1. One set (18 containers) was kept in moist sand boxes and placed under a transparent plastic cover and exposed to sunlight. Another set was kept in the dark to mimic the effect of a full macrophyte cover.

Table 1 Randomised block design for pH treatments under batch incubations

BLOCK 1			BLOCK 2		
$10*_C$	5_A	7_C	RWW_C	9_C	4_C
5_B	$RWW\#_B$	4_B	10_B	7_B	9_B
7_A	4_A	9_A	5_C	10_A	RWW_A

Numbers represent pH treatment and letters represent replications
 # RWW -raw wastewater without pH adjustment

The pH in each container was adjusted daily with 0.2N HCl or 0.2N NaOH. Temperature and dissolved oxygen concentration were taken twice daily at 8 and 14 GMT and data were averaged. Faecal coliform counts were determined every other day for 15 days.

Results

Environmental conditions

In all ponds, the highest temperature (31-34°C) was reached in the middle of the day (13GMT). Temperature at 8GMT was between 25 and 27°C. At 20 GMT, it was between 29 and 31°C. There was a tendency of decreasing temperatures at 20GMT with increasing depth (Table 2, 3, 4 and 5).

The pH in the anaerobic pond was neutral at all depths at all times (Table 2). The pH in all 3 pond systems varied considerably (Table 3, 4 and 5). The water lettuce ponds became more acidic from pond 1 to 4. All duckweed ponds remained neutral at all times, while the algal pond system became more alkaline from pond 1 to 4. Only the algal ponds showed pH variation in depth and with time. High pH and strong pH fluctuations were observed at the surface of the algal ponds with the strongest fluctuations occurring in pond 4 (Table 5).

Table 2 Environmental conditions in the anaerobic pond

Depth	Time (in GMT)			Time (in GMT)		
	8	13	20	8	13	20
	Temperature (^{0}C)			DO (mg/L)		
Surface (0.10m)	24.7±1.3	32.1±0.8	30.3±3.5	0.5±0.2	0.3±0.2	0.3±0.3
Middle (0.35m)	24.7±1.2	30.7±3.7	29.9±2.5	0.4±0.2	0.3±0.3	0.3±0.3
Bottom (0.63m)	24.7±1.1	30.6±3.0	29.1±3.0	0.4±0.2	0.3±0.2	0.3±0.2
	pH			TDS (mg/L)		
Surface (0.10m)	7.1±0.7	7.0±0.5	7.2±0.1	589±57	548±90	556±10
Middle (0.35m)	7.1±0.4	7.0±0.5	7.1±0.0	587±56	547±70	557±8
Bottom (0.63m)	7.1±0.4	6.9±0.5	7.1±0.1	590±74	555±30	559±9

±: *standard deviation*

After passing through the anaerobic pond the water was nearly anaerobic (DO is 0.3-0.5mg/L) and TDS was between 555 and 590mg/L (Table 2). The DO in the water lettuce and duckweed ponds increased gradually from pond 1 to pond 4 (Table 3 and 4). However, DO in the algal ponds showed a totally different behaviour. There was an immediate increase in pond 1 with the strongest effect at 13GMT, which led to a depth gradient (Table 5).

The decrease in TDS was much more pronounced in the water lettuce ponds (about 100mg/L in pond 4) than in the duckweed and algal ponds (about 300mg/L in pond 4 for both systems) (Table 3, 4 and 5).

Table 3 Environmental conditions in the water lettuce pond system

Pond	Depth	Time (in GMT)			Time (in GMT)		
		8	13	20	8	13	20
		Temperature (^{0}C)			DO (mg/L)		
1	Surface (0.10m)	26.4±1.2	32.2±1.2	31.0±1.9	0.9±0.5	0.5±0.4	1.4±0.7
	Middle (0.35m)	26.8±1.0	32.4±1.0	30.2±1.6	0.5±0.1	0.4±0.2	1.1±0.6
	Bottom(0.63m)	27.0±1.7	32.0±1.7	29.8±1.5	0.5±0.2	0.5±0.3	0.6±0.4
2	Surface(0.10m)	26.4±1.2	31.7±1.2	31.0±1.7	0.5±0.8	0.9±0.5	1.3±0.4
	Middle(0.35m)	26.8±0.9	32.2±0.9	30.2±1.5	0.1±0.7	0.6±0.2	1.1±0.2
	Bottom (0.63m)	27.0±1.7	31.5±1.7	29.9±1.4	0.2±0.5	0.5±0.3	1.0±0.8
3	Surface (0.10m)	26.2±1.2	31.4±1.2	30.9±1.7	1.6±0.7	3.2±0.9	3.0±2.0
	Middle(0.35m)	26.5±1.1	32.1±1.3	30.1±1.5	1.3±0.5	2.2±0.2	2.1±0.4
	Bottom(0.63m)	27.1±2.0	31.7±2.0	29.8±1.5	1.0±0.8	2.4±1.1	1.7±0.4
4	Surface (0.10m)	26.1±1.1	31.9±1.0	30.8±1.8	2.5±1.1	4.3±2.7	4.4±1.5
	Middle(0.35m)	26.8±1.1	32.9±1.1	30.2±1.4	2.6±1.0	3.9±2.1	4.4±1.5
	Bottom (0.63m)	26.9±1.3	31.6±1.3	29.6±1.4	2.1±0.8	3.3±2.5	1.5±2.0
		pH			TDS (mg/L)		
1	Surface(0.10m)	6.3±0.2	6.3±0.2	6.3±0.3	263± 97	259±101	245±98
	Middle(0.35m)	6.3±0.2	6.4±0.3	6.3±0.3	266±95	260±97	245±98
	Bottom(0.63m)	6.3±0.2	6.4±0.2	6.4±0.2	268±93	260±95	246±98
2	Surface(0.10m)	5.7±0.5	5.7±0.6	5.6±0.9	135±60	133±60	129±58
	Middle(0.35m)	5.6±0.6	5.7±0.6	5.6±0.8	137±61	134±57	130±56
	Bottom(0.63m)	5.7±0.5	6.1±0.2	5.6±0.9	137±60	133±58	130±59
3	Surface(0.10m)	4.8±0.4	4.8±0.4	4.9±0.4	90±3	90±2	89±2
	Middle (0.35m)	4.8±0.5	4.8±0.4	5.0±0.4	91±2	92±1	89±2
	Bottom(0.63m)	4.8±0.6	4.9±0.5	5.0±0.5	92±3	91±1	90±2
4	Surface (0.10m)	4.3±0.0	4.4±0.1	4.5±0.3	96±1	95±2	98±2
	Middle (0.35m)	4.3±0.0	4.4±0.1	4.6±0.4	95±2	96±3	97±3
	Bottom (0.63m)	4.4±0.2	4.3±0.0	4.5±0.4	94±1	95±2	97±4

±: *standard deviation*

Table 4 Environmental conditions in the duckweed pond system

Pond	Depth	Time (in GMT) 8	13	20	Time (in GMT) 8	13	20
		Temperature (^{0}C)			DO (mg/L)		
1	Surface (0.10m)	27.3±0.9	33.1±0.9	29.8±1.5	0.7±0.7	0.4±0.2	1.7±0.9
	Middle (0.35m)	27.8±0.3	32.6±0.3	28.4±1.1	0.8±0.6	0.6±0.3	1.4±0.5
	Bottom (0.63m)	28.2±1.1	32.3±1.1	28.1±1.2	0.6±0.4	0.6±0.4	1.0±1.1
2	Surface (0.10m)	27.1±1.2	32.8±1.2	29.7±1.3	2.2±1.8	0.7±0.4	1.5±0.1
	Middle (0.35m)	27.8±0.3	32.5±0.3	28.3±0.9	1.6±0.7	0.6±0.4	1.4±0.2
	Bottom (0.63m)	27.9±0.8	32.2±0.8	27.9±0.8	0.8±0.8	0.8±0.9	0.5±0.3
3	Surface (0.10m)	26.7±1.5	32.7±1.5	29.7±1.5	2.8±0.6	1.5±0.5	2.6±0.3
	Middle (0.35m)	27.7±0.5	32.7±0.5	28.4±1.0	2.4±0.5	1.2±0.3	2.1±0.8
	Bottom (0.63m)	27.8±1.3	31.9±1.3	28.1±0.8	0.8±0.9	0.7±0.7	1.0±0.3
4	Surface.(0.10m)	26.9±1.0	32.5±1.0	29.7±1.3	3.4±1.1	2.4±0.9	3.1±1.1
	Middle (0.35m)	27.7±0.6	32.9±0.6	28.2±0.9	3.6±1.0	2.2±0.9	2.6±1.6
	Bottom (0.63m)	27.7±1.4	31.8±1.4	28.1±1.0	2.4±1.5	1.5±1.1	2.8±0.9
		pH			TDS (mg/L)		
1	Surface (0.10m)	6.9±0.2	7.0±0.0	7.0±0.0	557±70	530±56	559±52
	Middle (0.35m)	6.9±0.1	7.0±0.1	7.1±0.0	551±179	532±174	564±184
	Bottom (0.63m)	6.0±0.1	7.0±0.0	7.1±0.0	555±170	539±165	574±174
2	Surface (0.10m)	7.0±0.2	6.9±0.0	7.0±0.0	462±59	421±46	449±150
	Middle (0.35m)	7.0±0.1	6.9±0.1	7.1±0.1	449±147	425±137	458±159
	Bottom (0.63m)	7.0±0.1	7.0±0.1	7.1±0.1	452±139	433±131	470±145
3	Surface (0.10m)	6.8±0.2	6.8±0.1	6.9±0.1	355±69	339±65	363±44
	Middle (0.35m)	6.8±0.2	6.8±0.1	7.0±.0.1	351±113	340±108	367±118
	Bottom(0.63m)	6.8±0.2	6.8±0.1	7.0±.0.1	350±110	344±106	375±114
4	Surface (0.10m)	6.6±0.3	6.7±0.1	6.8±0.1	261±64	257±67	265±59
	Middle(0.35m)	6.7±0.3	6.8±0.2	6.8±0.1	262±85	258±84	266±85
	Bottom (0.63m)	6.7±0.3	6.7±0.2	6.8±0.1	261±84	260±83	266±84

Table 5 Environmental conditions in the algal pond system

Pond	Depth	Time (in GMT)			Time (in GMT)		
		8	13	20	8	13	20
		Temperature (^{0}C)			DO (mg/L)		
1	Surface (0.10m)	26.1±1.0	34.2±1.0	29.9±1.2	0.8±0.2	15.3±0.8	2.3±1.8
	Middle (0.35m)	26.7±0.5	32.8±0.5	28.7±0.9	0.9±0.9	2.4±2.1	1.8±1.4
	Bottom (0.63m)	27.1±1.0	32.4±1.0	28.5±0.9	0.4±0.1	1.1±1.4	0.7±0.7
2	Surface (0.10m)	26.0±0.9	34.4±0.9	29.9±1.4	1.3±1.1	14.3±2.2	2.1±0.9
	Middle (0.35m)	27.1±0.7	33.3±0.7	28.8±1.1	0.7±0.2	2.1±0.1	2.0±1.0
	Bottom (0.63m)	27.0±1.2	32.1±1.2	28.7±1.1	0.6±0.3	1.0±1.1	0.6±0.4
3	Surface (0.10m)	26.1±1.1	34.3±1.1	29.8±1.5	1.0±0.3	13.5±4.6	1.9±1.4
	Middle (0.35m)	27.0±0.8	33.9±0.8	29.0±1.2	0.6±0.2	2.5±1.2	1.2±0.9
	Bottom (0.63m)	26.8±1.2	33.2±1.2	28.7±1.4	0.6±0.2	1.5±1.8	0.6±0.4
4	Surface (0.10m)	26.0±1.0	34.8±1.0	30.0±1.4	3.5±1.0	17.1±0.9	10.5±1.0
	Middle (0.35m)	26.7±1.1	34.2±1.1	29.2±1.2	2.4±1.0	5.3±3.1	8.2±0.1
	Bottom (0.63m)	27.0±1.0	33.6±1.0	29.0±1.2	1.2±0.7	3.1±3.2	2.1±1.4
		pH			TDS (mg/L)		
1	Surface (0.10m)	7.4±0.4	8.5±0.8	7.5±0.4	433±76	368±10	417±66
	Middle (0.35m)	7.5±0.4	7.7±0.4	7.5±0.3	423±136	383±111	424±135
	Bottom (0.63m)	7.3±0.2	7.6±0.3	7.4±0.2	420±71	413±66	428±65
2	Surface (0.10m)	7.4±0.1	8.6±1.2	7.5±0.3	377±46	298±16	349±46
	Middle (0.35m)	7.3±0.3	7.5±0.2	7.4±0.3	370 ±122	319±104	354±113
	Bottom (0.63m)	7.3±0.3	7.4±0.1	7.4±0.3	370±52	358±50	356±48
3	Surface (0.10m)	7.2±0.2	8.4±1.1	7.1±0.4	316±15	272±18	273±53
	Middle (0.35m)	7.0±0.0	8.1±0.6	7.1±0.3	318±108	295±95	300±89
	Bottom (0.63m)	7.0±0.0	6.9±0.3	7.0±0.2	323±26	297±27	306±30
4	Surface (0.10m)	7.8±0.9	10.0±0.1	9.3±1.0	276±90	314±9	279±19
	Middle (0.35m)	8.1±1.0	9.6±0.5	9.0±0.8	275±95	275±104	273±92
	Bottom (0.63m)	7.8±0.7	8.9±0.4	8.4±1.0	274±92	274±10	286±11

± : *standard deviation*

Faecal coliform removal

The faecal coliforms in all pond systems decreased from pond 1 to pond 4. There was not much difference between the duckweed and the algal treatment systems in the first three ponds in the amount of faecal coliforms removed. A remarkable difference between the treatment systems was observed in the last pond. The log removal of faecal coliforms in the systems was 2 for water lettuce, 3 for duckweed and 5 for algal pond systems (Figure 2 and Table 6).

Table 6 Log removal of faecal coliform in macrophyte and algal based waste stabilization ponds

Pond System	Pond 1	Pond 2	Pond 3	Pond 4
Water lettuce	1.0	1.5	1.6	2.0
Duckweed	2.0	2.5	2.8	3.0
Algae	2.0	2.5	3.0	5.0

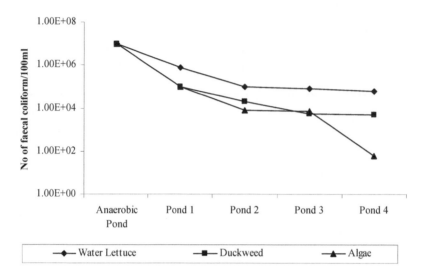

Figure 2 Removal of faecal coliforms in macrophyte and algal ponds

Effect of pH on faecal coliform removal rate on batch scale incubations

Environmental conditions in batch scale pH experiments.
Algal colonization only occurred in the raw wastewater exposed to sunlight on the 2nd day after start of experiment of batch scale incubations and on the 5th day for pH treatments 7 and 9 exposed to sunlight. The raw wastewater exposed to sunlight had pH values in the range of 7-10.5 and DO values of >10mg/l due to algae colonization.
DO concentrations of the pH treatments exposed to sunlight were higher than those in darkness (Figure 3).

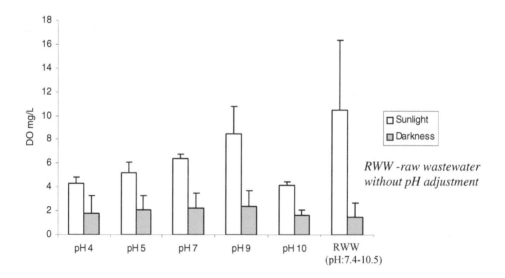

Figure 3 DO concentrations in different pH treatments under sunlight and in darkness

Faecal coliform removal

Faecal coliform removal rates were generally higher in sunlight than in darkness but were not significantly different apart from pH 4 treatments (Table 7). The removal rates were not significantly ($p<0.05$) different for pH 4, 7, 9 and 10 in sunlight and pH 9, 10 and raw treatments in darkness. The least detrimental effects on faecal coliforms were observed at pH 5 (Table 7).

Table 7 Die-off rate of faecal coliforms under sunlight and dark conditions

pH treatment	Removal rate (d^{-1})	
	Sunlight	Darkness
4	1.5a*	1.1c
5	0.2b	0.4b
7	1.3a	1.0c
9	1.4a	1.2ac
10	1.4a	1.4a
raw#	1.4a	1.2ac

 # Raw wastewater without pH adjustment   *
Different letters mean significant differences (p< 0.05)

Discussion

Environmental conditions

Temperatures in all ponds were within 25-34^{0}C, an optimal range for indicator bacteria and also suitable for wastewater treatment (Reed *et al.*, 1988). Mayo and Noike (1996) and Jagger (1985) observed that within this temperature range there were no significant effects on bacteria growth, that is to say that the growth rate virtually remains the same.

The acidic conditions in the water lettuce ponds are in line with the findings of Sharma and Sridhar (1985) and Awuah *et al* (2001; 2002). Kansiime and van Bruggen (2001) also found acidic conditions in wetlands in with *Cyperus papyrus* and *Miscanthidium* receiving domestic wastewater. Contrarily, Sooknah and Wilkie (2004) found pH 8 in water lettuce and water hyacinth ponds. The acid conditions in our ponds could be caused by the release of CO_2 into the wastewater by water lettuce plants during respiration through the roots and from heterotrophic bacteria activity. This means that photosynthesis of water lettuce plants may not consume CO_2 from the water but directly from the atmosphere. Duckweed on the other hand may consume CO_2 from the water via its roots as well and thus they create a neutral condition. Algae can only use dissolved CO_2 and therefore create high pH conditions during photosynthesis.

According to (Baumgartner, 1996) the DO in water is caused by gas-liquid mass transfer at the surface and subsequent mixing throughout the depth of the water and oxygen released during photosynthesis minus the oxygen consumed during respiration. The algal ponds were directly exposed to sunlight this led to high DO concentrations from air diffusion and increased photosynthetic oxygen production of algae. The amount used in respiration appears lower than the amount obtained from diffusion and photosynthesis. Low DO concentrations in the macrophyte ponds could be explained by reduced oxygen diffusion from the atmosphere into the water column due to the plant cover, higher respiration rates and oxygen uptake by microorganisms attached to the roots and photosynthetic oxygen release directly into the atmosphere. The first and second water lettuce and duckweed ponds did not show differences in

DO concentrations. The last two water lettuce ponds had higher DO concentrations than the last two duckweed ponds. This is contrary to what Moorhead and Reddy (1988) found in their study to evaluate oxygen transport into the root zone for several floating and emergent aquatic macrophytes. They found transport rates of oxygen were highest in macrophytes with small root mass. Sooknah and Wilkie (2004) also observed highest DO concentrations in ponds covered by pennywort, which had smaller root mass compared to water lettuce and water hyacinth ponds and concluded that small root mass is associated with high DO levels in macrophyte ponds.

The total dissolved solids were evenly distributed at all depths, which indicate that solutes are well mixed. Awuah et al. (2003) observed that the active zones in all 3 pond systems were over 70% which indicates a good mixing of pond constituents. Water lettuce ponds removed more dissolved solids than the duckweed fronds with smaller root mass. Similarly, Sooknah and Wilkie, (2004), found a correlation between root mass and removal of TDS. The greater the quantity of root mass the better the removal of TDS.

The water lettuce plants having more extensive root systems were able to absorb more nutrients for growth than the duckweed fronds and this may be the reason for higher production of biomass found in water lettuce ponds compared to duckweed ponds (Awuah et al., unpublished).

Faecal coliform removal

The differences in the removal of faecal coliforms in the 3 pond systems could be due to several factors. In the algal ponds high pH and DO levels as well as the pH fluctuations could be the main factors for faecal coliform removal (Curtis et al., 1992; Awuah et al., 2004). The faecal coliform removals correlated well with high pH and DO concentrations observed in the algal ponds. Davies-Colley et al. (1999) and Pearson et al. (1987) have attributed pathogen removal in pond systems to several factors such as direct sunlight, high pH levels and DO concentrations, which are mainly found in the last algal pond. Faecal coliform removal in this study occurred mainly in the last algal pond where pH levels and DO concentrations were the highest (Table 5). High pH levels and DO concentrations appeared at the surface of the algal ponds; hence any wastewater passing at the surface could have an efficient removal of faecal coliforms. This may necessitate the positioning of effluent outlet pipes for algal ponds near the surface to enhance faecal coliform removal.

Low pH and DO levels were observed in the water lettuce ponds and neutral conditions and low DO levels were observed in the duckweed ponds. Although low removal of faecal coliforms was observed in the macrophyte ponds, the faecal bacteria profile study in Awuah et al., 2004 (using the same low strength sewage) showed that at the middle level (0.35m) faecal coliform populations in the water lettuce ponds are comparable to that of the algal pond effluent. The water lettuce roots near the surface might have harboured faecal coliforms, thus increasing faecal coliform concentrations on the surface and subsequently carried into the effluent by water current. The word harboured is used meaning that the faecal coliforms were staying around the roots either as attached, detached or both, or living close for exudates etc from the root zone. The outlet pipes of the three treatment systems used in this study were placed near the surface. Placing outlet pipes of water lettuce ponds in the middle section of last pond may produce better effluent quality. Since high pH and DO events do not usually occur in macrophyte ponds, other factors such as natural die-off, long retention periods, attachment and predation could be responsible for faecal coliform removal.

Faecal coliform die-off rates at different pH on batch scale incubation

In the batch-sale studies, algae colonization occurred in the incubations under sunlight exposures for pH 7 and 9 but because pH was artificially stabilised, effect of increase in pH by algae might not really count. Algae colonized raw wastewater easily, which led to increase in pH and high DO concentrations. Towards the end of the experimental period, conditions of the raw wastewater were similar to pH 9 and 10. No algae colonization was observed at pH 4 and yet the removal rate in sunlight was significantly different in the treatment under darkness. This shows the importance of direct sunlight on faecal coliform die-off at pH 4. Curtis et al., (1992), explained that the damage to the cytoplasmic membranes of bacteria is caused by sunlight through DNA damage causing the formation of more photo products including pyrimidine dimers. These substances further cause cessation of growth and ultimately death (Liltved and Landfold, 2000). Many researchers (Polprasert et al. 1983; Gersberg et al. 1987; Curtis et al. 1994; Davies-Colley et al. 1999) have shown direct sunlight to be lethal to faecal coliforms. For the rest of the treatments, sunlight did not seem to have any significant impact on faecal coliform removal. The insignificant effect of sunlight on the faecal coliforms removal observed in this study showed that, the faecal coliforms are probably used to sunlight exposures or able to photo repair all damages caused by sunlight under such pH treatments. It is also possible that the presence of algae might have diminished the effect of sunlight. Other factors appeared to be playing a role here especially in the raw wastewater treatments where there were no significant difference in removal between exposures under sunlight and in the dark in spite of the high pH and DO concentrations under sunlight conditions. It is however possible that the high pH and the high DO concentrations might have affected faecal coliform removal at pH 9 and 10.

Troussellier et al., (1986) reported that high pH was the key factor controlling faecal bacteria die-off in sewage lagoons. Pearson et al., (1987) also reported that faecal coliforms appeared to be adversely affected by high pH and DO. The possible reason given by Zepp et al., (1981) and Haag et al., (1986) was that high pH increases the production of toxic forms of oxygen which may damage the cytoplasmic membrane leading to the death of the bacteria. It is also possible that the hydroxyl ions damage the cell membranes and cause death. The highest die-off rate occurred at pH 4 in sunlight with a low DO concentration. There was also low DO concentration at pH 10 in sunlight. Raw wastewater in sunlight had high DO above saturation levels. The die-off rates at pH 4, 10 and the raw wastewater (pH 7-10.5) under sunlight exposures were not different from each other. This suggests that pH effects may be more important in the removal of faecal coliforms than DO concentrations. The low die-off rates observed at pH 5 cannot be explained. Also Vincent et al. (1994) gave no explanation for the same observation.

The die-rates of faecal coliforms in this study are comparable to that of enterococci under similar conditions (Awuah et al., 2001). However, the die-off rates for enterococci were higher than that of faecal coliforms. The behaviours at different pHs were however not the same. At pH 4 for example, enterococci die-off was not affected by sunlight even though in both cases the highest die-off rate was observed at this pH in sunlight. Faecal coliforms die-off rate was significantly detrimental in the presence of sunlight at pH 4. The enterococci were also more susceptible to sunlight and dissolved oxygen than faecal coliforms used in this study.

Conclusions

Acidic conditions exist in water lettuce ponds, neutral conditions in duckweed ponds and alkaline conditions in algal ponds. No significant difference in die-off rates for faecal coliforms was observed at pH 5, 7, 9 and 10 under both sunlight and dark exposures. pH 5 is least detrimental to faecal coliforms. High pH is more effective in the removal of faecal coliforms than DO. Extremes of pH (4, 9 and 10) could be the most important factor in the removal of faecal coliforms in stabilization ponds.

Acknowledgement

This research was made possible by grants from the Netherlands government through the SAIL foundation.

References

Awuah E, Asante K, Anohene F, Lubberding HJ and Gijzen HJ (2001). Environmental conditions in macrophyte and algal domestic wastewater treatment systems. *Wat. Sci. Tech.* **44** (6), 11-18.

Awuah E, Lubberding HJ, Asante K and Gijzen HJ (2002). The effect of pH in the removal of pathogens in Pistia, duckweed and algal-based stabilization ponds for domestic wastewater treatment. *Wat. Sci. Tech.* **45**(1), 67-74.

Awuah E, Oduro-Kwarteng S, Lubberding HJ and Gijzen HJ (2003). Hydraulic behaviour of macrophyte and algal wastewater treatment systems. In: *KNUST SERR 2 Proceedings.* Elmina, Ghana, August 2002. 261p.

Awuah E, Oppong-Peprah M, Lubberding HJ. and Gijzen HJ. (2004). Comparative performance studies of macrophyte and algal-based stabilization ponds. *J. Toxicol. Env. Health* Part A, **67**, 1-13.

Baumgartner D J (1996). Surface water pollution. In: Pepper JL, Gerba PC and Brusseau ML (eds). Pollution Science. Academic Press Inc. Canada, pp. 189 – 220.

Curtis TP, Mara DD and Silva SA (1992). Influence of pH, oxygen, and humic substances on ability of sunlight to damage faecal coliforms in waste stabilization pond water. *Appl. Environ. Microbiol.* **58**, 1335-1343.

Davies-Colley RJ, Donnison AM, Speed DJ, Ross CM and Nagels JW (1999). Inactivation of faecal indicator microorganisms in waste stabilization ponds: Interactions of environmental factors and sunlight. *Wat. Res.* **33**, 1220-1230.

Feacham RG, Bradley DJ, Garelick H and Mara DD. (1983). Sanitation and disease: Health aspects of excreta and wastewater management. Wiley, UK.

Gersberg RM, Lyon SR, Brenner R, and Elkins BV (1987). Fate of viruses in artificial wetlands. Appl. *Environ. Microbiol.* **83**, 731-736.

Gilmour CC. (1992). Effects of acid deposition on microbial processes in natural waters. In: Mitchell, R (ed). Environmental microbiology. Wiley-Liss Inc. New York, pp. 33-57.

Greenberg AE, Clesceri LS and Eaton AD (1992). Standard methods for examination of water and wastewater. 18th edition, American Public Health Association. American Water Works Association. Water Environmental Fed. Washington. D.C.

Haag WR, Hoigne J, Gassman E and Brunn A (1986). Singlet oxygen in surface water, Part III. Photochemical formation and steady-state concentrations in various types of water. *Chemosphere* **13**, 641-650.

Jagger J (1985). Solar –UV radiations on Living cells. Praeger, New York, pp. 174-176.

Kansiime F and van Bruggen JJA (2001). Distribution and retention of faecal coliforms in the Nakivobu wetland in Kampala, Uganda. *Wat. Sci. Tech.*. **44**(6), 199-206.

Liltved H and Landfold B (2000). Effects of high intensity light on ultraviolet –irradiated and non irradiated fish pathogenic bacteria. *Wat. Res.* **34**, 81-486.

Maynard HE, Ouki SK and Williams SC. (1999). Tertiary Lagoons: A review of removal mechanisms and performance. *Wat. Res.* **33**, 1-13.

Moorhead KK, Reddy KR, 1988. Oxygen transport through selected aquatic macrophytes. *J. Environ. Qual.* **17**, 138–142.

Mayo AW and Noike T (1996). Effects of temperature and pH on the growth of heterotrophic bacteria in waste stabilization ponds. *Wat. Res.* **30**, 447-455.

Pearson HW, Mara DD, Mills SW and Smallman DJ (1987). Physicochemical parameters influencing faecal bacteria survival in waste stabilization ponds. *Wat. Sci. Tech.* **19**(12), 145-152.

Polprasert C, Dissanayake MG and Thanh MC (1983). Bacterial die-off kinetics in waste stabilization ponds. *Wat. Pollut. Contr. Fed.* **55**, 285-296.

Reed SC Middlebrooks EJ and Crite, RH (1988). Natural systems for waste management and treatment. McGraw-Hill, USA, 433p.

Sooknah RD and Wilkie AC (2004). Nutrient removal by floating aquatic macrophytes cultured in anaerobically digested flushed dairy manure wastewater. Ecol. Eng. 22, 27–42.

Sharma BM and Sridhar MKC (1985). Some observations on the oxygen changes in a lake covered with Pistia stratiotes. L. *Wat Res.* **19**, 953-939.

Troussellier M, Legendre P and Baleux B (1986). Modelling of the evolution of bacterial densities in an eutrophic ecosystem (sewage lagoons). Microbial Ecol. **12**, 355-379.

Vincent G, Dallaire S and Lauzer D (1994). Antimicrobial properties of root exudates of three macrophytes: Mentha aquatica L., Phragmites australis (CAV) Trin. and Scirpus lacustris L In Wetland systems for Water Pollution Control Conf. Proc. ICWS Secretariat Guangzhou P.R. China, pp 290-296.

Von Sperling M (1996). Comparison among the most frequently used systems for wastewater treatment in developing countries. *Wat. Sci. Tech.* **33** (3), 59-76.

Wetzel RG (2001). Limnology of lake and river ecosystems. 3rd Edition Academic Press. San Diego, 429p.

Zepp RG, Baughman GL and Schlotzhauer PF (1981). Comparison of photo-chemical behaviour of various humic substances in water. II Photosynthesised Oxygenations. *Chemosphere* **10**, 119-126.

Chapter Five

Effect of pH Fluctuations on Pathogenic Bacteria Removal in Domestic Wastewater

Awuah E, Lubberding HJ and Gijzen HJ
Effect of pH fluctuations on pathogenic bacteria removal in domestic wastewater

Submitted to *Water Research*

Effect of pH Fluctuations on Pathogenic Bacteria Removal in Domestic Wastewater

Abstract

Fluctuations in pH may occur in waste stabilization ponds. This behaviour could be bactericidal. However, no studies have been conducted to confirm this. The present study addresses the issue by studying the die-off of *Escherichia coli*, coliforms, *Salmonella* and other enterobacteria in domestic wastewater. Fluctuating pH conditions were established in incubations with pH values ranging between 4-9, 7-8, 7-9, 7-10, and 7-11, while stable pH incubations were at pH 4, 5, 7, 8, 9, 10 and 11. A follow up on the experiment was done in sterilised domestic wastewater using *E. coli* (ATCC13706) and enterococci isolated from domestic wastewater. All experiments were done in replicates of three in 100ml plastic containers. Adjustment of pH was done twice a day at 08GMT and 16GMT with 0.2N HCl or 0.2N NaOH. Stable pH appeared more detrimental than fluctuating pH to *E. coli* and coliforms in domestic wastewater and *E. coli* (ATCC 13706), except fluctuating pH of 4-9, which had higher die-off rates than stable pH of 4, and 9 for *E. coli*. Fluctuations in pH had higher die-off rates than stable pH incubations for *Salmonella* and other enterobacteria. At pH 5, low die-off rates were recorded for *E. coli* and *Salmonella*. Increase in numbers was observed for *E. coli* (ATCC 13706), coliforms and other enterobacteria in at pH 5 and 8 incubations respectively. For enterococci, fluctuations in pH were more detrimental than stable pH incubations. Extreme pH treatment of 4, 10, and 11 were found to be most detrimental to all the faecal bacteria used in this study. Lower die-off rates were observed in sterilized wastewater incubations than non-sterilized wastewater incubations. This study shows that pH fluctuations, extremes of pH and presence of other microbes may all contribute to pathogen die-off in domestic wastewater.

Key words: Pathogens; pH; fluctuations; domestic wastewater

Introduction

Several hypotheses have tried to explain the causes of pathogen indicator reduction in natural wastewater treatment systems. Some of the mechanisms studied include sunlight and sunlight triggered mechanisms such as high pH and DO levels common in algal waste stabilization ponds (Parhad and Rao, 1972; 1974; Whitlam and Codd, 1988; Curtis *et al.*, 1992), the depletion of nutrients (Portier and Palmer, 1989), protozoa grazing (Gersberg *et al.*, 1987), attachment to macrophytes (Spira *et al.*, 1981) and the presence of antibacterial substances produced by algae (Mezrioui and Oudra, 1998) and macrophytes (Vincent *et al.*, 1994).

However, most emphasis was given to pH effects on indicator bacteria (Parhad and Rao, 1972; 1974, Pearson *et al.*, 1987, Curtis *et al.*, 1992; 1994; Davies-Colley *et al.*, 1999).
Most indicator organisms can grow only within the pH range of 4-9 (Rheinheimer, 1992). Prescott *et al.*, (1996), however, raised the minimum pH requirement for indicator growth to 5.5 and the upper limit to pH 8.

Several researches conducted on pH effects on indicator organisms (Pearson *et al.*, 1987; Curtis *et al.*, 1992; 1994; Davies-Colley *et al.*, 1999) have been done within the neutral and alkaline ranges. Diurnal fluctuations of pH levels accompanied by high faecal coliform removal in algal ponds have also been reported (Pearson *et al.*, 1987; Awuah *et al.*, 2001; 2002; Zimmo *et al.*, 2003; Awuah *et al.*, 2004). Feachem *et al.*, (1983) also reported that a large fluctuation of pH ranging from 7 and 10 could be detrimental to bacteria and subsequently change the composition of bacterial populations. Faecal coliform removal in waste stabilization ponds may be due to these pH fluctuations. In earlier studies, we found low pH (4) in water lettuce ponds and high pH (>9) with fluctuations in diurnal pH in algal ponds. Low pH was also found to be detrimental to enterococci (Awuah *et al.*, 2002, 2004). It is important to know, if the fluctuations in pH in waste stabilizations ponds could lead to a better removal of faecal bacteria. The understanding of such pH removal mechanisms in ponds could contribute to an improvement in operation and management practices of stabilization ponds to create conditions, which are detrimental to pathogens found in wastewater. The aim of this study is to answer the question: Do pH fluctuations affect faecal bacteria removal more than stable pH?

Materials and Methods

An experiment was conducted to determine the effect of pH fluctuations on *Escherichia coli*, coliforms, *Salmonella* and other enterobacteria in raw domestic wastewater. The pH fluctuations used were in the ranges: 4-9, 7-8, 7-9, 7-10, and 7-11. Besides, the following stable pH were used: 4, 5, 7, 8, 9, 10, 11 and raw wastewater without any pH adjustment was also included. In addition, the effect of pH fluctuations on *E. coli* (ATCC 13706) and enterococci was studied in sterilised domestic wastewater. The experiments were conducted in replicates of three in 100mL opaque plastic containers using domestic wastewater from the Kwame Nkrumah University of Science and Technology (KNUST) campus in Kumasi, Ghana. The wastewater used was medium strength sewage with BOD between 260-320mg/L and NH_4^+-N between 32.4 to 48.8mg/L (Awuah *et al.*, 2004). The pH in the containers was adjusted twice daily at 8GMT and at 16GMT using 0.2N HCl or 0.2N NaOH. The presence of *E. coli*, coliforms, *Salmonella*, and other enterobacteria was determined on chromocult agar after incubation at 37^0C for 24hrs (Byamukama *et al.*, 2000). The medium is able to differentiate between *E. coli*, *Salmonella*, other coliforms (coliforms other than *E. coli*) and other enterocbacteria, which could not be differentiated by pigmentation. Confirmatory tests were conducted for *E. coli* using EC medium (Greenberg *et al.*, 1992). The confirmatory tests for *Salmonella* were done using growth in tetrathionate base broth as a positive result. Coliforms were confirmed by acid and gas production in MacConkey broth purple at 35^0C for 24 to 48 hrs and metallic colonies on Endo agar (Greenberg *et al.*, 1992). Enterococci were determined in Slanetz Bartley medium (Niemi and Ahtiainen, 1995). Confirmatory tests were done by Gram reaction for Gram-positive cocci in chains and growth in 6.5% saline broth (Greenberg *et al.*, 1992).
The average numbers of bacteria from the various dilutions and replications were used to plot a graph based on Chick's law and the decay rates were calculated.

Results

Effect of pH fluctuation on faecal bacteria in domestic wastewater

The behaviour of E. coli was similar to that of coliforms in general. Results for the two categories of bacteria showed that stable pH causes faster bacterial die-off than pH fluctuations. For E. coli at low pH and neutral pH, fluctuations were more effective than stable pH except for pH 7-8. At higher pH there was no difference for pH treatment at 7-11, 10 and 11. Fluctuated pH incubations of 4-9 and 7-9 had higher die-off rates than their stable pH counterparts. E. coli was completely eliminated at pH 11 treatments in just a few days (Table 1).

Table 1 Removal rates of faecal bacteria in domestic wastewater at different pH incubations

Escherichia coli				Other coliforms			
PH range	k (d^{-1})	stable pH	k (d^{-1})	pH range	k (d^{-1})	stable pH	k (d^{-1})
4-9	2.7a[1]	4	1.6f	4-9	2.7a	4	0.8c
7-8	0.4b	5	0.4b	7-8	1. 4b	5	0.6c
7-9	2.0c	7	1.5f	7-9	0.7c	7	0.3d
7-10	2.5a	8	1.5f	7-10	1.7b	8	+0.8e
7-11	3.3d	9	1.7f	7-11	1.5b	9	2.1f
raw*	1.1e	10	3.4d	Raw	0. 7c	10	1.6b
		11	3.2d			11	1.9f
Salmonella				Other enterobacteria			
PH range	k (d^{-1})	stable pH	k (d^{-1})	pH range	k (d^{-1})	stable pH	k (d^{-1})
4-9	1.2a	4	1.6f	4-9	0.0a	4	0.0a
7-8	1.0a	5	0.3g	7-8	0.6b	5	+0.1a
7-9	0.5b	7	1.4f	7-9	2.5c	7	0.0a
7-10	3.0ce	8	0.2g	7-10	2.7c	8	0.4d
7-11	3.5d	9	0.9a	7-11	0.6b	9	0.5b
raw	0.7b	10	2.8e	Raw	0.2e	10	0.6b
		11	3.2c			11	0.8b

*raw: raw wastewater (pH range 6.8-7.5) [1]Different letters mean significant differences (p< 0.05)

In the case of *Salmonella*, pH fluctuations of 7-10 and 7-11 were more bactericidal than stable pH incubations of 10 and 11. For other enterobacteria, all fluctuations were more detrimental than stable pH except at pH 11 and corresponding fluctuation treatment pH 7-11. pH 4 and corresponding fluctuation treatment pH 4-9 which showed no difference in removal rates. Increase in bacteria numbers occurred creating positive kd values for coliforms at pH 8 and for other enterobacteria at pH 5 incubations. It is of interest to note that at stable pH 11 *Salmonella* persisted after *E. coli* had been completely eliminated.

Effect of pH on E. coli (ATCC 13706) and enterococci in sterilised domestic wastewater
Almost all stable pH incubations were more detrimental than fluctuating pH incubations with the exception of stable pH 5. At pH 11, all the *E. coli* (ATCC 13706) bacteria were eliminated within a few hours in all three replicates. It must be noted that this *E. coli* (ATCC13706) was a pure viral free culture. The previous *E. coli* was isolated from domestic wastewater. The removal rate at stable pH of 4 was the highest among the treatments. Increase in bacteria numbers were observed at pH 5 and fluctuating incubations of 7-8 and 7-9 (Table 2).

Table 2 Removal rates of *Escherichia coli* (ATCC1307) and enterococci in sterilized domestic wastewater at different pH incubations.

Escherichia coli (ATCC13706)				Enterococci			
PH range	k (d^{-1})	stable pH	k (d^{-1})	pH range	k (d^{-1})	stable pH	k (d^{-1})
4-9	0.4a[1]	4	4.3d	4-9	0.6a	4	1.1c
5-7	1.4b	5	+0.1c	5-7	1.4b	5	1.1c
7-8	+0.4c	7	0.1e	7-8	1.0c	7	0.8c
7-9	+0.3c	8	0.4a	7-9	1.2b	8	0.9c
7-10	0.6a	9	0.0e	7-10	1.0c	9	0.9c
7-11	#	10	1.5b	7-11	0.8c	10	0.9c
Raw*	0.3a	11	#	Raw	0.1d	11	0.9c

*Raw: raw wastewater (pH range 6.8-7.5) # Eliminated in a few hours +: increase in numbers
[1]Different letters mean significant differences (p< 0.05)

Removal rates of *E. coli* in non-sterilised wastewater were higher than in sterilised wastewater. The removal rates at neutral pH in sterilised wastewater were very low.

For enterococci, removal rates in treatments with fluctuating pH were higher than stable pH treatments with the exception of pH 4, which had a higher removal rate than that of fluctuating pH treatment 4-9. Removal rates obtained for all stable pH treatments were almost the same. There were also no differences between pH treatment at 10, 11 and the fluctuated treatments of pH 7-10 and pH 7-11 incubations (Table 2). For enterococci all removal rates did not differ significantly, except 5-7 and 7-9 that were somewhat higher and 4-9 that was lower than corresponding stable pH treatments. In general, k did not change too much (from 0.6 until 1.4).

Discussion

There were some similarities in the behaviour for *E. coli*, *Salmonella* and coliforms because they all belong to the same family Enterobacteriaceae with evolutionary adaptation to the intestinal tract. All enterobacteria studied showed more resistance to acid than alkaline conditions. Stable pH at extremes of 10 and 11 appeared to be more detrimental to *E. coli* and coliforms than pH fluctuating treatments. *Salmonella* and other enterobacteria were however, more susceptible to pH fluctuations than stable pH treatments. *E. coli* was more susceptible to high pH 11 than the other faecal bacteria. They were eliminated in a few hours for *E. coli* (ATCC13706) in sterilized sewage and in a few days for *E. coli* in raw sewage. This means that in wastewater treatment plants if *E. coli* alone is used as an indicator for treatment efficiencies, it might not be able to guarantee the absence of pathogens like *Salmonella* at high pH, which are associated with algal ponds. We recommend the use of other indicator organisms apart from *E. coli* in algal pond systems to evaluate the performance of wastewater treatment plants to avoid situations like this. The die-off rates for other coliforms were lower than that of *E. coli*, and *Salmonella*. The die-off rates of other enterobacteria are also comparable to low die-off rate observed for faecal coliforms in chapter 4 of this thesis. This shows that the total faecal coliform test could be better than using *E. coli* alone as a pathogen indicator. The use of chromocult agar, which can distinguish between four categories of bacteria including *E. coli*, *Salmonella*, other coliforms and other enterobacteria is appropriate for the detection of the presence of both pathogens and indicator organisms.

Alkaline conditions were more detrimental than acidic conditions for *E. coli*, *Salmonella*, coliforms and other enterobacteria with exception of pH 4. *E. coli* and *Salmonella* are known to have stressed tolerance mechanisms which are induced under acidic conditions; they possess acid shock proteins (Foster and Moreno, 1999), which can help the organism to survive at pH 2 (Lin *et al.*, 1995). However, in our studies the acid tolerance mechanism was not observed in these strains of faecal bacteria at pH 4.

The low removal rates of faecal bacteria and increase in faecal bacteria numbers observed at pH 5 are a confirmation of earlier works by Awuah *et al.*, (2003). Vincent *et al.*, (1994) also observed that at pH 5 there were no significant removal of *E. coli* in comparison to the effects of pH 4 and 3.8 which caused significant removal of *E. coli*, but they could not give any reason for this behaviour of *E. coli*. This behaviour of *E. coli* at pH 5 is contradictory to the findings of Prescott *et al.*, (1996) and Solic and Kortulovic (1992) who reported that at pH 5 a sharp decline in bacteria numbers including *E. coli* could occur.

High pH of 10 and especially 11 had strong effects on all faecal bacteria. Free ammonia (NH_3) is high at these pHs. Free ammonia is toxic for all bacteria including *E. coli* (Deal *et al.*, 1975).

Most studies have found *E. coli* to be highly susceptible to high pH associated with stabilization ponds (Pahard and Rao, 1974; Pearson *et al.*, 1987; Awuah, *et al.*, 2003). Curtis *et al.*, (1994) attributed this susceptibility to visible sunlight, UV light and photooxidation due to reactive oxygen species released in algal ponds in the presence of sunlight. They suggested that the active oxygen species have a relation with sunlight and high pH. The toxicity of ammonia released at alkaline pH was not discussed in their paper. The pH was also not controlled in their studies. The wastewater in our study had negligible DO levels (Awuah *et al.*, 2004) and direct sunlight effects were avoided by conducting the experiments in the laboratory using opaque containers under 40watts for short durations to avoid algal colonization. Thus, the removal at high pH could be due to ammonia toxicity rather than sunlight and DO concentrations. Although ammonia concentrations were not measured in this study, they are expected to be high. The total ammonia concentrations found in the wastewater used was between 32 and 48mg/L (Awuah *et al.*, 2004), which is above inhibitory concentrations of 25-30mg/L (Strauss *et al.*, 1997). It is possible that in stabilization ponds with high pH, ammonia toxicity could be a major contributing factor to the elimination of faecal coliforms in addition to the photooxidation process (Curtis *et al.*, 1994). This will require further investigations.

Stable, high pH (10, 11) is more detrimental than fluctuating pH (7-10, 7-11) levels. Also, the high removal rates at pH fluctuation in the range of 4-9 for *E. coli* and coliforms showed promising application of alternating water lettuce (pH 4) and algal ponds (high pH) under low organic loadings. Such wide pH ranges are detrimental to faecal bacteria (Feacham *et al.*, 1983). However, this may not be true for all faecal bacteria as seen in these studies for *Salmonella*, other enterobacteria and enterococci. Alternation of macrophyte and algal ponds may achieve similar results. The added advantage will be the elimination of suspended algae especially if the last pond is macrophyte-based which blocks sunlight penetration into the water column. This may help to prevent blockage by algae in sprinklers during re-use for drip irrigation. Whether the removal of faecal bacteria will be enhanced or not when macrophyte and algal ponds are alternated, will require additional investigation in pilot and full scale systems.

Increase in *E. coli,* coliforms and other enterobacteria as observed for some treatments in this study may be due to bacteria growth, but the possible release of attached bacteria (contributing to higher numbers) can not be excluded. Solo-Gabriele *et al.*, (2000) also found *E. coli* multiplication in soil as a major source of contamination in coastal waters. Faecal coliform growth in water bodies has also been reported in tropical regions (Santiago-Mercado and Hazen; 1987; Rivera *et al.*, 1988; Byappanahalli and Fujioka, 2004).

Enterococci were affected more by fluctuations in pH than stable pH except at stable pH 4 treatment. The treatments at stable pH however had almost the same die-off rates. The changes in die-off rates were not much for all stable pH treatments, though effect of alkaline conditions appeared least detrimental while acidic conditions appeared more detrimental. The mechanism associated with the low tolerance to acidic conditions and tolerance to alkaline pH by enterococci however, is poorly understood. Moellering, (1995) also reported of some tolerance of enterococci to high pH.

Removal rates in non-sterilized domestic wastewater were higher than that of the sterilized wastewater incubations. The release of nutrients during sterilization may promote growth (Awuah and Lorbeer, 1991). The destructions of other organisms and viruses (bacteriophages) in sewage through sterilisation might have eliminated competition for nutrients, predation and bacterial lysis caused by viruses (Maynard *et al.*, 1999).

The results showed that faecal coliform removal in algal ponds is caused by high pH generated and the removal of faecal bacteria in water lettuce ponds is caused by low pH. In the duckweed ponds extreme pH conditions are absent and other factors such as attachment, predation, sedimentation and long retention periods could be important in pathogen removal. These mechanisms could also be important in water lettuce and algal ponds.

Conclusion

It appears that under different pH fluctuation incubations, faecal bacteria behave differently depending on species. Fluctuations in pH can be more detrimental than stable pH depending on species. Acidic conditions tend to enhance enterococci die-off more than alkaline conditions, while acidic conditions appear less bactericidal to *E. coli*, coliforms, *Salmonella* and other enterobacteria especially at pH 5. Low pH of 4 and high pH values of 10 and 11 can be detrimental to faecal bacteria. These extremes of pH may account for pathogen removal in pond systems. This study shows that pH fluctuations, extremes of pH, high free ammonia concentration under high pH, and presence of other microbes may all contribute to die-off in stabilization ponds. The high pH and DO conditions associated with algal ponds are absent in macrophyte ponds and mechanisms other than direct sunlight penetration and pH induced factors may be important in the removal of pathogens in macrophyte ponds. These may include attachment, predation, sedimentation and long retention periods.

Acknowledgement
This research is made possible by grants from the Netherlands government through the SAIL foundation.

References

Awuah RT and Lorbeer JW (1991). Methyl bromide and steam treatment of an organic soil for control of fusarium yellows of celery. *Plant disease* **75**, 123-136.

Awuah E, Asante K, Anohene F, Lubberding HJ and Gijzen HJ (2001). Environmental conditions in macrophyte and algal-based domestic wastewater treatment systems. *Wat. Sci. Tech.* **44**(6), 11-18.

Awuah E, Asante K, Lubberding, HJ and Gijzen HJ (2002). The effect of pH on enterococci in *Pistia,* duckweed and algal-based stabilization ponds for domestic wastewater treatment. *Wat. Sci. Tech.* **45** (1), 67-74.

Awuah E, Boateng J, Lubberding HJ and Gijzen HJ (2003). Physico-chemical parameters and their effects on pathogens in domestic wastewater. In: *KNUST SERR 2 Proceedings.* Elmina, September, 2002. 261p.

Awuah E, Oppong-Peprah M, Lubberding HJ and Gijzen HJ (2004). Comparative performance studies of macrophyte and algal-based stabilization ponds. *J. Toxicol. and Environ. Health.* Part A, **67**, 1-13.

Byamukama D, Kansiime F, Mach RL and Farnleitner H (2000). Determination of *Escherichia coli* contamination with chromocult coliform agar showed a high level of discrimination efficiency for differing faecal pollution levels in tropical waters of Kampala, Uganda. *Appl. Environ. Microbiol.* **66,** 864-868.

Byappanahalli M and Fujioka R (2004). Indigenous soil bacteria and low moisture may limit but allow faecal bacteria to multiply and become a minor population in topical soil. *Wat. Sci. Tech.* **50**(1), 27-32.

Curtis TP, Mara DD and Silva SA (1992) Influence of pH, oxygen, and humic substances on ability of sunlight to damage faecal coliforms in waste stabilization ponds. *Appl. Environ. Microbiol.* **58**, 1335-1343.

Curtis TP, Mara DD, Dixo NGH and Silva SA (1994). Light penetration in waste stabilization ponds. *Wat. Res.* **28**, 1031-1038.

Davies-Colley RJ, Donnison AM, Speed DJ, Ross, CM and Nagels JW (1999). Inactivation of faecal indicator microorganisms in waste stabilization ponds: Interactions of environmental factors and sunlight. *Wat. Res.* **33**, 1220-1230.

Deal PH, Souza, KA and Mack HM (1975). High pH ammonia toxicity, and the search for life on the Jovian planets. *Orig. Life.* **6**(4), 561-73.

Facklam RR, Sahm DS and Teixeira LM (1999). Manual of clinical Microbiology. Internet site: http://www.enterococcus.ouhsc.edu/lab_methods.asp.

Feachem RG, Bradley DJ, Garelick H and Mara DD (1983). "Sanitation and disease: health aspects of excreta and wastewater management". World Bank studies in Water Supply and Sanitation.

Foster JW and Moreno M (1999). Inducible tolerance in enteric bacteria. *Norvatis Found. Symp.* **221**, 55-69.

Gersberg R, Brenner R, Lyon S and Elkins B (1987). Survival of bacteria and viruses in municipal wastewater applied to artificial wetlands. In: Reddy K and Smith W (eds). *Aquatic plants for wastewater treatment using artificial wetlands.* Magnolia Publishing, Orlando Florida. pp. 237-245.

Greenberg AE, Clesceri LS and Eaton AD (1992). Standard methods for the examination of water and wastewater. APHA/AWWA. Water Environmental Fed. Washington DC.

Lin J, Lee IS, Frey J, Slonczewski JL and Foster JW (1995). Comparative analysis of extreme acid survival of *Salmonella typhimurium. J. Bacteriol.* **177**, 4097-4104.

Maynard HE, Ouki SK and Williams SC (1999). Tertiary Lagoons: A review of removal mechanisms and performance. *Wat. Res.* **33**, 1-13.

Mezrioui NE and Oudra B (1998). Dynamics of picoplankton and picroplankton flora in the experimental wastewater stabilization ponds of the arid region of Marrakech, Morocco and cyanobacteria effects on *Escherichia coli* and *Vibrio cholerae* survival. In: Wong YS and Tam NFY (eds). *Wastewater treatment with algae.* Springer-Verlag and Landes Bioscience Publishers 234p.

Mitchell R (1992). Environmental microbiology. Wiley-Liss Inc. New York, 411p.

Moellering RCJ (1995). *Enterococcus* species, *Streptococcus bovis*, and *Leuconostoc* species. In: Mandell GL, Bennett JE and Dolin R (eds). *Principles and Practices of Infectious Diseases,* 4th ed. Churchill Livingston, New York, pp. 1826-1835.

Niemi RM and Ahtiainen J (1995). Enumeration of intestinal enterococci and interfering organisms with Slanetz-Bartley agar, KF streptococcus agar and the MUST method. *Lett. Appl. Microbiol.* **20**(2), 92-7.

Parhad NM and Rao NU (1972). The effect of algal growth on the survival of *E coli* in sewage. *Indian J. Environ. Health.* **14**, 131-139.

Parhad NM and Rao NU (1974). Effect of pH on survival of *E. coli. Wat. Pollut. Contr. Fed.* **46**, 149-161.

Pearson HW, Mara DD, Mills SW and Smallman DJ (1987). Physicochemical parameters influencing faecal bacteria survival in waste stabilization ponds. *Wat. Sci. Tech.* **19**(12), 145-152.

Portier R and Palmer S (1989). Wetlands microbiology: form, function, process. In: Hammer DA (ed.) *Constructed wetlands for wastewater treatment: municipal, industrial and agricultural*. Lewis publishers Chelsea, Michigan, pp. 89-106.

Prescott LM, Harley JP and Klein DA (1996). Microbiology, 3rd edition, W.C. Brown Publishers, 558p.

Rheinheimer G (1992), "Aquatic microbiology", 4th Edition. John Wiley and Sons, Chichester. 184p.

Santiago-Mercado J and Hazen TC (1987). Comparison of four membrane filter methods for fecal coliform enumeration in tropical waters. *Appl. Environ. Microbiol.* **53**, 2922- 2928.

Singh A and Mcfeters GA (1992). Detection methods for water borne pathogens. In: Mitchell R (ed.). *Environmental Microbiology*. Wiley Liss Inc. New York, pp. 125-126.

Solo-Gabriele HM, Wolfert MA, Desmarais, TR and Palmer CJ (2000). Sources of *Escherichia coli* in a coastal subtropical environment. *Appl. Environ. Microbiol.* **66**, 230-237.

Solic M and Kortulovic N (1992). Separate and combined effects of solar radiation, temperature, salinity and pH on the survival of faecal coliforms in seawater. *Marine Pollution Bulletin* **24**, 411-416.

Spira WM, Hug A, Ahmed QS and Saeed YA (1981). Uptake of *Vibrio cholerae* biotype eltor from contaminated water by water hyacinth (*Eichhornia crassipes*). *Appl. Environ. Microbiol.* **42,** 50-553.

Strauss M, Larmie SA and Heinss U (1997). Treatment of sludges from on-site sanitation low-cost options. *Wat. Sci. Tech.* **35** (6), 129-136.

Rivera SC, Hazen TC and Toranzos GA (1988). Isolation of fecal coliforms from pristine sites in a tropical rain forest. *Appl. Environ. Microbiol.* **54**, 513-517.

Vincent G, Dallaire S and Lauzer D (1994). Antimicrobial properties of root exudates of three macrophytes: *Mentha aquatica* L., *Phragmites australis* (CAV) Trin. and *Scirpus lacustris* L. In: *Wetland systems for Water Pollution Control Conf. Proc.* ICWS Secretariat Guangzhou P.R. China, pp. 290-296.

Whitlam GC and Codd GA (1988). Damaging effects of light on microorganisms, *Spec. Publ. Soc. Gen. Microbiol.* **17**, 129-169.

Zimmo O (2003). Nitrogen transformations and removal mechanisms in algal and duckweed waste stabilization ponds. PhD dissertation. UNESCO-IHE, Delft Netherlands, 131p.

Chapter Six

Comparative Performance Studies of Water Lettuce, Duckweed and Algal Stabilization Ponds Using Low Strength Sewage

Adapted from

Awuah E, Oppong-Peprah M, Lubberding HJ and Gijzen HJ (2004)

Comparative performance studies of water lettuce, duckweed and algal-based stabilization ponds using low strength sewage

Journal of Toxicology and Environmental Health, Part A **67**, 1-13

Comparative Performance Studies of Water Lettuce, Duckweed and Algal-Based Stabilization Ponds Using Low Strength Sewage

Abstract

A bench scale continuous flow wastewater treatment system comprising of 3 parallel lines using water lettuce (*Pistia stratiotes*), duckweed (*Spirodela polyrhiza*) and algae (natural colonization) as treatment agents was set up to determine environmental conditions, faecal coliform profiles and general treatment performance. Each treatment system consisted of 4 ponds connected in series and was fed with diluted sewage. Influent and effluent parameters measured included environmental conditions, turbidity, biochemical oxygen demand (BOD), chemical oxygen demand (COD), nitrate, nitrite, ammonia, total phosphorous, faecal coliforms, mosquito larvae and sludge accumulation. Environmental conditions and faecal coliform profiles were determined at the surface (0.10m), middle (0.35m) and bottom part (0.63m), of each pond. Acidic conditions were observed in the water lettuce ponds, neutral conditions in duckweed ponds and alkaline conditions in algal ponds. Faecal coliform log removal of 3, 4 and 6 were observed in water lettuce, duckweed and algal pond systems respectively in the final effluents, with log removal rate per pond of 1.6, 2.0 and 2.7 respectively. Sedimentation accounted for over 99% faecal coliforms removal in most of the water lettuce and algal pond systems. BOD removal was least in the algal system, followed by water lettuce and duckweed at 25%, 93% and 95% respectively. COD removal was 59% and 65% respectively for water lettuce and duckweed whiles it increased in algal ponds by 56%. Total phosphorus removal was 9% for duckweed systems and 33% for water lettuce pond system while it increased by 19% in the algal pond system. Ammonia removal was 93% for the algal and 95% in both water lettuce and duckweed pond systems. Removal of total dissolved solids (TDS) were 9%, 15% and 70% respectively for algae, duckweed and water lettuce. Mosquito populations of $96/m^2$, $3,516/m^2$ and $11,175/m^2$, and were found in duckweed, algal and water lettuce ponds respectively.

Keywords: Water lettuce; duckweed; algae; stabilization ponds; performance

Introduction

Domestic wastewater is increasingly becoming a problem in many developing countries as urban centres grow. Waste stabilization ponds are known to be effective in treating wastewater (Landsdell, 1987; Wang, 1991). They are cheap to construct and are effective in pathogen removal (Feacham et al., 1983; Von Sperling, 1996) making it safe for reuse in agriculture. In spite of the advantages of stabilization ponds, they do not usually provide sufficient incentives to support maintenance cost. The use of macrophyte ponds has been recommended as an alternative to algal ponds (Gijzen and Khonker, 1997) with the possibility of resource recovery as an incentive, but the performance in developing countries where this technology will be most appropriate has not been thoroughly investigated. Also, it is unclear how macrophyte systems compare to algal ponds with respect to main treatment parameters.

This study was therefore conducted in Ghana, a tropical developing country, using water lettuce (*Pistia stratiotes)* and duckweed (*Spirodela polyrhiza*), to determine their general performance with particular reference to pathogen removal. In previous research conducted on a batch scale, the performance of the macrophyte ponds was found to be efficient in terms of pathogen and nutrient removal with no significant differences between the water lettuce, duckweed and algal ponds (Awuah *et al.*, 2001). Environmental conditions were however different with acidic conditions characterizing the water lettuce ponds, neutral in duckweed ponds while alkaline conditions characterized the algal ponds, (Awuah *et al.*, 2001). Since environmental conditions and pathogen die-off on a batch scale may be different in comparison to continuous flow field conditions, a similar experiment was conducted using a continuous flow system.

Materials and Methods

Wastewater from the Kwame Nkrumah University of Science and Technology (KNUST) in Kumasi, Ghana was used. The wastewater was diluted by a ratio of 1:3, sewage to tap water and put into an anaerobic pond for 2 days. After the anaerobic treatment, the wastewater was distributed over three pilot scale stabilization pond treatment lines, one with water lettuce, one with a duckweed cover and another with algae (natural colonization). Each treatment system consisted of a series of 4 ponds (volume 80L m^3 each) with a retention period of 7 days per pond. A flow rate of 0.01m^3/day was maintained in all the pond systems under gravity, resulting in a total retention time of 28 days each treatment system (Figure1). Dense coverage by water lettuce and duckweed was maintained. Water lettuce was partially harvested once a week and duckweed twice a week.

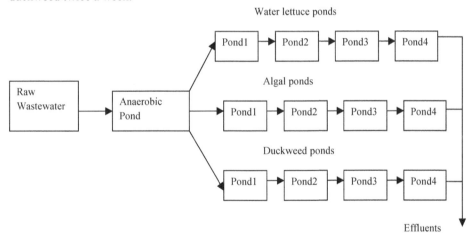

Figure 1 Schematic diagram of pilot scale macrophyte and algal waste stabilization ponds

The study was carried out one month after algae colonization had appeared, which was visible from the green coloration and lasted from February to August, 2000. After 6 months of operation, the containers were decanted and the sludge accumulations were measured. Periodically, mosquito larvae populations were measured by scooping from the surface with 1L containers (of surface area 86.6cm^2). This was left in the ponds for a few minutes. Larvae were identified and counted. pH, dissolved oxygen (DO), total dissolved solids (TDS) and temperature were determined at different levels (0.1m), (0.35m) and (0.63m) of each pond of the three pond systems at 8, 13 and 20 h Greenwich meridian time (GMT). Faecal coliforms were determined in samples taken at 8 GMT at various depths (0.1m), 0.35m) and 0.63m). They were measured once a week for 8 weeks using membrane lauryl sulphate broth on membrane filters. Environmental conditions, i.e. temperature, pH, DO and TDS were measured in situ 4 times during the studies with WTW 323-B/set electronic probes. Suspended solids, biochemical oxygen demand (BOD), chemical oxygen demand (COD), phosphate, ammonia nitrogen, nitrite-nitrogen and nitrate-nitrogen were measured at influent and all effluent points each week for 8 weeks using the Standard Methods in Greenberg *et al.*, (1992). Turbidity was measured with Nephelometric turbidity meter. Data analyses were conducted based on the students' "t" test and standard deviation.

Results

Temperature, pH and DO for the raw wastewater and anaerobic pond were almost the same. The average faecal coliform concentration of the raw wastewater was $3.5 \times 10^8/100\text{ml}$ and that of the effluent from the anaerobic pond was $2.8 \times 10^7/100\text{ml}$. The BOD and COD characteristics of the raw sewage showed that the sewage was above the medium strength (Metcalf and Eddy, 2003). The raw wastewater had an average BOD of 285 mg/L and COD of 696mg/L. Ammonia nitrogen was 38mg/L. The concentrations in the anaerobic pond effluent were 80mg/L for BOD, 182mg/L for COD and 18.66mg/L for ammonia (Table 1).

Table 1 Characteristics of raw wastewater and anaerobic pond effluent

Parameter	Raw wastewater	Anaerobic effluent
Temperature (°C)	28.8±1.8	30.7±0.7
DO (mg/L)	0.5±0.1	0.3±0.2
pH	7.8±0.5	7.4±0.1
TDS mg/L	1052 ±353	397.3±228.8
Faecal coliforms/100mL	$3.5 \times 10^8 \pm 4.0 \times 10^8$	$2.8 \times 10^7 \pm 1.2 \times 10^7$
COD (mg/L)	696.0±28.8	182.0±52
BOD (mg/L)	284.7±25.3	80±1.5
NO_3-N (mg/L)	4.9±2.1	1.9±0.7
NO_2-N (mg/L)	0.1±0.1	0.5±0.9
NH_3-N mg/L	38.1±8.2	18.7±2.5
Total phosphorous (mg/L)	15.3±6.3	4.3±2.3

The environmental conditions presented in Table 2 are a combination of diurnal and profile readings. Thus, the values appear low in comparison to afternoon values, which are normally higher (Awuah *et al.*, 2001). Environmental conditions measured showed that the water lettuce ponds had acidic conditions and low oxygen levels. Duckweed ponds had neutral pH and low DO levels. Algal ponds were alkaline (>pH 8) and had high levels of oxygen (>5mg/L) in the last pond. The pH and DO in the last ponds of the algal pond system were higher than in the two macrophyte systems with significant (P<0.05) differences (Table 2). The total dissolved solids were least in the water lettuce ponds followed by duckweed and algal systems in the last pond (Table 2).

Table 2   Average environmental conditions in macrophyte and algal ponds
(Diurnal and profiles combined)

Pond systems		Temperature (°C)	pH	DO (mg/L)	TDS (mg/L)
Anaerobic (effluent)		30.7a*	7.4d	0.3a	397.3a
Water lettuce	Pond 1	29.8a	6.3a	0.7b	257.0e
	Pond 2	29.6 a	5.7a	0.8b	166.2b
	Pond 3	29.5 a	4.9b	2.1c	90.6c
	Pond 4	29.6 a	4.4b	3.3d	95.9c
Duckweed	Pond 1	29.7 a	6.9c	0.9b	551.2a
	Pond 2	29.3 a	7.0c	1.1b	446.4a
	Pond 3	29.5 a	6.8c	1.7c	353.8d
	Pond 4	29.5 a	6.7c	2.7d	261.6e
Algae	Pond 1	29.6 a	7.6d	2.7d	412.2a
	Pond 2	29.7 a	7.5d	2.8d	349.9d
	Pond 3	29.9 a	7.3d	2.6d	299.9d
	Pond 4	30.1 a	8.8e	5.9e	280.7d

Numbers in the same column with the same letter showed no significant differences (P<0.05)

The log removal of faecal coliforms per pond was in an increasing order of 1.6 for water lettuce, 2.0 for duckweed and 2.7 for algae in each treatment system (Table 3). Faecal coliform levels declined progressively along the series. Significant differences (P<0.05) between pond systems were observed in the last ponds only (Figure 2).

Lowest faecal coliform counts were observed at the surface in the algal and duckweed ponds and in the middle of the water column in the water lettuce ponds. Most of the faecal coliforms were found in the sediments of all pond systems (Table 3; Figure 3). The log_{10} removal from the first pond to the last ponds in the sediment layer was zero for duckweed and 3 for algae and water lettuce (Table 3).

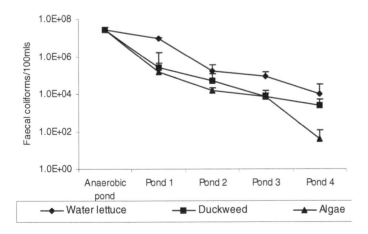

Figure 2 Faecal coliform removal in macrophyte and algal ponds

Samples from the middle of the last pond in the water lettuce treatment system showed highest faecal coliform removal with log removal of 4, while it was 3 for algae and duckweed. In the surface profiles, log removal from the first pond was 2 for water lettuce, 2 for duckweed and 4 for algae. From the anaerobic pond to the final effluent, log removal of 3, 4 and 6 were observed in water lettuce, duckweed and algal pond systems respectively (Table 3). Increasing retention period, high pH and DO were associated with high faecal coliform removal (Tables 2 and 3).

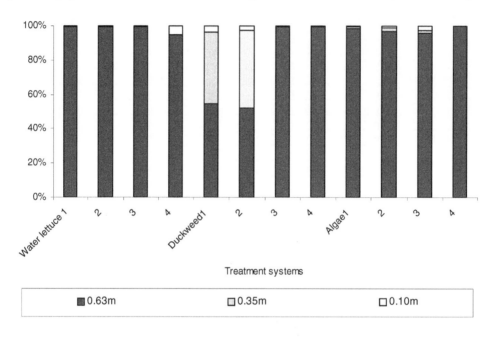

Figure 3 Faecal coliform profiles in macrophyte and algal ponds in percentages

Table 3 Faecal coliform profiles and die-off in pond systems

	Ponds	0.1m	0.35m	0.63m	k
	1	9.6E+5±8.8E+5	1.3E+5±1.4E+5	5.1E+7±6.1E+7	
	2	1.7E+5±2.0E+5	6.0E+4±6.4E+4	4.0E+7±5.0E+7	
	3	8.3E+4±7.3E+4	9.0E+3±6.9E+3	7.8E+6±1.3E+7	
Water lettuce	4	9.6E+3±2.2E+4	7.0E+1±9.0E+1	2.3E+5±2.6E+5	1.6
		0.1m	**0.35m**	**0.63m**	**k**
	1	2.5E+5±2.1E+5	2.3E+6±3.3E+6	3.1E+6±2.1E+6	
	2	5.3E+4±6.5E+4	9.7E+5±8.1E+5	1.2E+6±1.3E+6	
	3	8.3E+3±7.7E+3	1.1E+3±9.2E+2	2.8E+6±2.1E+6	
Duckweed	4	2.8E+3±2.9E+3	4.1E+3±4.7E+3	2.1E+6±2.4E+6	2.0
		0.1m	**0.35m**	**0.63m**	**k**
	1	1.2E+5±1.4E+5	1.2E+5±8.3E+4	1.5E+7±2.5E+7	
	2	1.6E+3±6.3E+3	3.9E+4±2.6E+4	1.6E+7±1.8E+6	
	3	7.5E+3±3.6E+3	4.1E+3±1.7E+3	2.8E+5±1.2E+5	
Algae	4	5.0E+1±8.0E+1	2.7E+2±3.8E+2	1.9E+5±1.0E+5	2.7

k=Log removal of faecal coliforms/pond

The organic matter declined with time. The BOD: COD ratio also declined with time with the highest declines recorded in the macrophyte-based ponds (Table 4). The BOD removal was not significantly (P<0.05) different in the water lettuce and duckweed ponds (Table 4). Percentage BOD removal was highest in the duckweed ponds (Table 5), followed by water lettuce and algae at 95%, 93% and 25% respectively. COD levels increased in the algal ponds by 56% (Table 5). COD removal was 65% and 59% for duckweed and water lettuce respectively. Ammonia removal was 95% for both water lettuce and duckweed and 93% in the algal ponds with no significant (P<0.05) differences between pond systems. Nitrate concentrations increased in the algal based system by 79% and decreased by 16% and 21% respectively for water lettuce and duckweed treatment systems. Nitrite levels increased in macrophyte ponds but decreased in algal the pond systems. Total phosphorus removal was 9% and 32% in duckweed and water lettuce ponds. The concentrations in the algal ponds increased by 19% (Table 5).

Table 4 Average organic and nutrients levels and BOD/COD ratios in macrophyte and algal ponds

Treatment System		Parameters in mg/L						BOD/COD ratio
		BOD	COD	Total P	NO$_3$-N	NO$_2$-N	NH$_3$-N	
Raw sewage		284.7a*	696.0a	15.3a	4.9a	0.11a	38.1a	0.41
Anaerobic pond effluent		80b	182.0b	4.3b	1.9b	0.50b	18.7b	0.43
Water lettuce	Pond 1	37.8c	109.9c	4.7b	2.4b	0.03c	36.3a	0.30
	Pond 2	25.5d	85.3d	2.8c	2.0b	0.33d	22.0c	0.29
	Pond 3	16.9e	83.2d	4.9b	2.0b	0.75e	11.4d	0.20
	Pond 4	4.2f	74.0e	2.9c	1.6b	0.66e	3.5e	0.06
Duckweed	Pond 1	24.1d	137.0b	3.9d	2.2b	0.02c	39.4a	0.18
	Pond 2	16.5e	111.3c	4.2d	1.7b	0.12a	27.0f	0.14
	Pond 3	12.6e	73.6e	4.2d	1.6b	0.27d	15.7d	0.17
	Pond 4	5.8f	64.0e	3.9d	1.5b	0.47b	4.7e	0.09
Algae	Pond 1	46.2g	197.4b	5.2e	1.5b	0.17a	26.8f	0.23
	Pond 2	85.8b	262.3b	5.2e	2.6c	0.02c	14.1d	0.33
	Pond 3	54.4g	307.2f	4.7b	3.2c	0.25d	8.3g	0.18
	Pond 4	59.2g	283.5f	5.1e	3.4c	0.07c	3.4e	0.21

Numbers in the same column with the same letter showed no significant differences (P<0.05)

Table 5 Removal efficiencies of macrophyte and algal ponds.

Parameter	Water lettuce	Duckweed	Algae
TDS(%)	70	15	9
Turbidity(%)	95	92	56
Faecal coliform removal (log)	3	4	6
BOD(%)	93	95	25
COD (%)	59	65	56 increase
NO$_3$-N(%)	16	21	79 increase
NH$_3$-N (%)	95	95	93
Total P(%)	33	9	19 increase

Sludge accumulations were lower in the anaerobic and macrophyte ponds than in the algal ponds (Figure 4). Turbidity measurements were significantly (P<0.05) higher in the algae ponds than in the macrophyte ponds (Figure 5). The water lettuce ponds had mosquitoe larvae in all the ponds with no significant difference between them. Mosquitoe larvae were present in all the algal ponds too. Apart from the second pond, the duckweed ponds did have no mosquitoe larva. *Culex* species were the only mosquitoes found during the studies. Mosquito larvae populations of 96/m^2, 3,516/m^2 and 11,175/m^2, were counted in duckweed, algal and water lettuce ponds respectively (Figure 6).

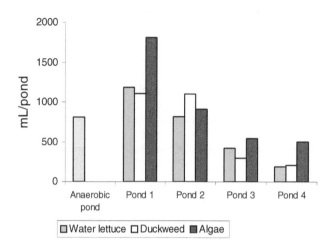

Figure 4 Sludge accumulations in macrophyte and algal ponds

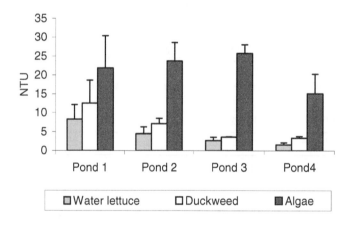

Figure 5 Turbidity in macrophyte and algal ponds

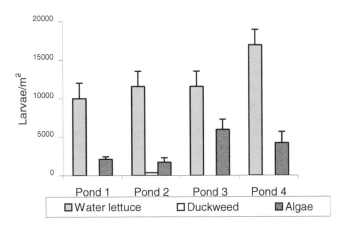

Figure 6 Mosquito larvae populations in macrophyte and algal ponds

Discussion

The environmental conditions observed were the same as the ones observed under the batch scale systems (Awuah *et al.*, 2001). Sharma and Sridhar (1985) also observed acidic conditions in water lettuce ponds in Nigeria. Attionu (1972) however, reported that alkaline conditions were associated with water lettuce ponds both under laboratory and field conditions in Ghana. The acidic conditions in the water lettuce ponds could be due to the inability of the water lettuce plants to use up all the CO_2 produced during respiration. The CO_2 then passes through the roots into the water. Atmospheric CO_2 is readily available for the leaves, which are exposed to the atmosphere. In the algal ponds, CO_2 is always a limiting factor and instead carbonates and bicarbonates present is utilized. However, this drives a reaction in equation 1-4 to produce more OH^- creating alkaline conditions found in algal ponds.

$$CO_2 + H_2O \longleftrightarrow H_2CO_3 \qquad (1)$$

$$H_2CO_3 \longleftrightarrow H^+ + HCO_3 \qquad (2)$$

$$HCO_3^- \longleftrightarrow H+ CO_3^{-2} \qquad (3)$$

$$CO_3^{-2} + H_2O \longleftrightarrow HCO_3^- + OH \qquad (4)$$

The removal of faecal coliforms is in conformity with that of Zimmo *et al.*, (2002) who reported that algal ponds are more efficient than duckweed in the removal of faecal coliforms. The removal of faecal coliforms in this study was probably due to several factors acting synergistically to bring about the changes observed. These included high pH (Curtis *et al.*, 1992; Parhad and Rao, 1972), temperature and dissolved oxygen (Pearson *et al.*, 1987), long retention period and sedimentation (Kansiime and van Bruggen, 2001), inactivation by UV light (Curtis *et al.*, 1992; Davies Colley *et al.*, 1999) exposure to biocides excreted by plants (Gersberg *et al.*, 1989), adsorption to organic matter (Bastein and Hammer, 1993) attack by lytic bacteria and viruses and protozoa grazing (Lijklema *et al.*, 1987; Brettar and Höfle, 1992). Faecal coliform removal is associated with high pH, high DO and sunlight effects (Awuah *et al.*, 2003; Davies-Colley *et al.*, 1999), all of which were more predominant in algal ponds than in macrophyte ponds.

The faecal coliform profiles observed in this study (Table 3; Figure 3) show that low faecal coliform populations might not always be encountered at the surface especially in water lettuce ponds. This is because the water lettuce roots provided a surface area for attachment in the surface region thus, harbouring high numbers of faecal coliforms. Faecal coliform concentrations in the middle of the water column in water lettuce ponds and that on the surface of the algal ponds had the lowest numbers. It is therefore recommended that ponds be designed such that the effluent outlet could be taken at levels with the lowest pathogen or faecal coliform counts.

The acidic conditions in the last water lettuce pond might have contributed to the removal of pathogens. Awuah *et al.*, (2003) showed that acidic conditions are detrimental to faecal coliforms. In the duckweed system, sediments seem to protect pathogens. The reason could be due to the neutral conditions, which characterized the duckweed ponds. Sedimentation played a major role in pathogen removal (Table 3 Figure 3) and factors like pond depth and shape, which affect sedimentation should be studied further with a view to enhance sedimentation in stabilization ponds. Kansiime and van Bruggen (2001) also found the highest levels of faecal coliforms in the sediment layers in some parts of a wetland dominated by *Cyperus papyrus* and *Miscanthidium violaceum*. Pond depth must be properly correlated with die-off rates to obtain a more holistic approach to pond design and engineering.

The die-off obtained in this study is comparable to that of enterococci in batch studies of macrophyte pond systems (Awuah *et al.*, 2001). The $\log_{10}$ removal of 6 observed in the algal ponds is one log higher than results by Yagoubi *et al.*, (2000) in wastewater stabilization ponds in Morocco using low strength sewage of COD of 200mg/L and BOD of 45mg/L. They found $\log_{10}$ removal ranging from 4 to 5.

The results obtained for faecal coliforms removal in this study does not conform to removal of enterococci in the same continuous flow system, which showed that enterococci removal was similar for the three systems (Awuah *et al.*, 2002). The organic matter removal was high in the macrophyte ponds and this was due to heterotrophic bacteria degradation without further addition of organic matter from algal biosynthesis. Algal growth in these ponds was limited due to shading by the macrophytes on the pond surface. The algal ponds produced new algal cells through photosynthesis that remained in suspension and substantially increased the organic matter content. This photoautotrophic growth can produce organic matter increases as high as 160-240% (Mayo, 1996). The organic matter removal in terms of COD for the duckweed ponds (65%) was low compared to results obtained by others (Zimmo *et al.*, (2002), Oron *et al.*, (1987), Nevertheless, removal efficiency was comparable to high energy input activated sludge performance in Nigeria with a similar climate where the removal of COD was 79% (Sangodoyin and Ugoji, 1993). The performance here also falls in the range reported by Maynard *et al.*, (1999) of 32-83 % as efficiency of treatment for BOD and COD of stabilization ponds in Africa.

Water lettuce probably absorbed ammonia or the ammonia could be converted into nitrites/nitrates and quickly absorbed by the plants. However, this was not investigated. The removal of ammonia in the duckweed ponds and algal ponds could be due to volatilization and absorption by duckweed fronds respectively (Bonomo *et al.*, 1997) and denitrification (Zimmo *et al.*, 2002). The high nitrate levels in the algal ponds are due to nitrification. Zimmo *et al.*, (2002) observed high nitrate levels in the algal ponds in comparison to duckweed ponds in Palestine. The removal of phosphorous was low in all 3 pond systems. This could be due to the fact that the plants were not healthy enough to absorb more of the phosphorous. Toms *et al.*, (1975) reported low phosphorous removal in tertiary lagoons where algal growth was low. The increase observed in phosphorous in algal ponds may be due to re-suspension of phosphorus in the algal biomass at high pH and DO (Mara and Pearson, 1986; Reed *et al.*, 1988).

The low turbidity was expected in the water lettuce and duckweed ponds. The macrophyte cover on the surface kept agitation low. The algae were however suspended and this gave a rather turbid environment. The high sludge accumulations in the algae pond were due to the fact that the algae were not harvested and eventually settled to the bottom as sediments. The macrophytes were harvested resulting in low sludge accumulations.

Mosquito breeding was common in the water lettuce and algal ponds. The water lettuce plants do not form a tight cover, thus leaving spaces for mosquitoes breeding. The morphology of water lettuce and the environmental conditions in the pond apparently provided an ideal haven for the larvae. According to Sainty and Jacobs, (1981) water lettuce harbours mosquito larvae. The duckweed ponds were virtually devoid of mosquito larvae. Duckweed fronds completely covered the surface of the ponds. According to Goma (1966), duckweed interferes with respiration and causes asphyxiation of mosquito larvae. Such respiratory interferences were absent in the water lettuce and algal ponds. All the species identified were *Culex* species, the vector of the filarial worm of elephantiasis, which is not prevalent in the area of study.

The performance in terms of pathogen, nutrient and organic matter removal showed significant differences in comparison to batch scale studies (Awuah *et al.*, 2001).

Conclusions

Algal ponds are more efficient in the removal of pathogens than macrophyte ponds, while macrophyte ponds are more efficient in organic matter and nutrient removal than algal ponds using low sewage strength. If effluent is discharged from the portion below the roots of water lettuce plants, pathogen removal could be comparable to algal ponds. Macrophytes may decrease the frequency of desludging due to low sludge accumulations. Duckweed may help in the substantial reduction of mosquitoes breeding in ponds. The benefit of macrophyte and algal waste stabilization ponds could be harnessed by combining both systems. Since this study was conducted by diluting sewage, the performance studies should be repeated for raw wastewater.

Acknowledgement

This study was supported with grants from the Netherlands Government through the SAIL Foundation.

References

Attionu RH (1976). Some effects of water lettuce (*Pistia stratiotes*) *Hydrobiologia* **50**, 245-256.

Awuah E, Asante K, Anohene F, Lubberding HJ and Gijzen HJ (2001) Environmental Conditions in Macrophyte and Algal-based domestic wastewater pond systems. *Wat. Sci. Tech.* **44**(6), 11-18.

Awuah E, Lubberding HJ, Asante K and Gijzen HJ (2002). The effect of pH on enterococci removal in *Pistia,* duckweed and algal-based stabilization ponds for domestic wastewater treatment. *Wat. Sci. Tech.* **45**(1), 67-74.

Awuah E, Boateng J, Lubberding HJ and Gijzen HJ (2003). Physico-chemical parameters and their effects on pathogens in domestic wastewater In: *KNUST SERR 2 Proceedings*. Elmina, September, 2002.

Bastein PK and Hammer DA (1993). The use of constructed wetlands for wastewater treatment and recycling. In: Moshiri GA (ed) *Constructed wetlands for water Quality improvement.* Lewis publishers, pp. 62-65.

Bonomo L, Pastorelli G and Zambon N (1997). Advantages and limitations of Duckweed-based wastewater pond systems. *Wat. Sci. Tech.* **35**(5), 239-246,

Brettar I and Höfle M (1992). Influence of ecosystematic factors on survival of *Escherichia coli* after large scale release into lake mesocosms. *Appl. Environ. Microbiol.* **58**,2201-2210.

Curtis TP Mara DD and Silva SA (1992). Influence of pH, oxygen and humic substances on ability of sunlight to damage faecal coliforms in waste stabilization ponds. *Appl. Environ. Microbiol.* **58**,1335-1343.

Davies-Colley RJ, Donnison AM, Speed DJ, Ross CM and Nagels JW (1999). Inactivation of faecal indicator microorganisms in waste stabilization ponds interactions of environmental factors with sunlight. *Wat. Res.* **33** (5), 1220-1230.

Feacham RG, Bradley DJ, Garelick H and Mara DD (1983). *Sanitation and Disease: Health Aspects of Excreta and Wastewater Management.* Wiley, UK.

Gersberg R, Lyon S and Elkins B (1989). Integrated wastewater treatment using artificial wetlands. In: Hammer DA (ed). *Constructed wetlands for wastewater.* Lewis publishers, Inc. Michigan, pp. 477-483.

Gijzen HJ and Khonker M (1997). An overview of the ecology, physiology, cultivation and application of duckweed. Lit. Review. Report no. 0896, World Bank, Duckweed Research Project. Bangladesh. .69p.

Goma LKH (1966). The Mosquito. Webb JE and Newell GE (eds), Hutchinson and Co. Ltd. UK. pp. 16-19, 48-62, 95-105.

Greenberg AE, Clesceri LS and Eaton AD (1992). Standard methods for the examination of water and wastewater. 18th edition, American Public Health Association. American Water Works Association. Water Environmental Fed. Washington. D.C.

Johnson DC, Enriquez CE, Pepper LI, Davies TO, Garba CP and Rose JB (1997). Survival of *Gardia, Cryptosporidium,* Polivirus and *Salmonella* in marine waters. *Wat. Sci. Tech.* **35**(2), 61-268.

Kansiime F and van Bruggen JJA (2001). Distribution and retention of faecal coliforms in the Nakivobu wetland in Kampala, Uganda. *Wat. Sci. Tech..* **44**(6), 199-206.

Landsdell M (1987). The development of lagoons in Venezuela. *Wat. Sci. Tech.* **19**(12), 55-60.

Lijklema L, Habekotte B, Hooijmans CM, Aalderink RH and Havelaar AH (1987). Survival of indicator organisms in a detention pond receiving combined sewer overflow. *Wat. Sci. Tech.* **19**(5/6), 547-555.

Mara DD and Pearson HW (1986). Artificial freshwater environments: waste stabilization ponds. In: Schoernborn, W. (ed) *Biotechnology,* Weinheim, Germany: VCH Verlagsgesellschaft, pp.177-206.

Maynard HE, Ouki SK and Williams SC (1999). Review paper, Tertiary lagoons: Review of removal mechanisms and performance. *Wat. Res.* **33**, 1-13.

Mayo AW (1996). BOD_5 removal in facultative ponds: Experience in Tanzania *Wat. Sci. Tech.* **34**(5/6), 447-455.

Metcalf and Eddy (2003). Wastewater Engineering; treatment, disposal and re-use. McGraw-Hill Inc. New York, 1334p.

Mills SW, Alabaster GP, Mara DD, Pearson HW, and Thitai WN (1992). Efficiency of faecal bacteria removal in waste stabilization ponds in Kenya. *Wat. Sci. Tech.* **26** (7/8), 1739-1748.

Oron G, deVegt A and Porath D (1987). The role of the operation regime in wastewater treatment with duckweed. *Wat. Sci. Tech.* **19**, (1) 97-105.

Parhad NM and Rao NU (1972). The effect of algal growth on the survival of *E. coli* in sewage. *Indian J. Environ. Health* **14**,131-139.

Pearson HW, Mara DD, Mills SW and Smallman DJ (1987). Physiochemical parameters influencing faecal bacteria survival in waste stabilization ponds. *Wat. Sci. Tech.* **19**(12), 145-152.

Reed SC, Middlebrooks EJ and Crites RH (1988). Post, NM and Halston, J (eds). *Natural Systems for Waste Management and Treatment*, McGraw-Hill, USA.

Sainty GR and Jacobs SWL (1981). Water plants of New South Wales. Water Resources Commission, New South Wales, 550p.

Saqqar MM and Pescod MB (1992). Modelling coliforms reduction in wastewater stabilization ponds. *Wat. Sci. Tech.* **26**(12), 1667-1677.

Sangodoyin AY and Ugoji C (1993). Coordinated wastewater treatment and reuse-a developing country's experience. *Int. J. Environ. Stud. Sect. A.* **43**, 281-286.

Sharma BM and Sridhar MKC (1985). Some observations on oxygen changes in lakes covered with *Pistia Stratiotes*. *Wat. Res.* **19**, 935-939.

Toms IP, Owens M and Hall JA (1975). Observations on the performance of polishing lagoons at a large regional works. *Wat. Pollut. Contr.* **74**, 383-401.

Van Buuren JCL and Hobma SWL (1981). Faecal coliforms die-off at post treatment of anaerobically pre-treated domestic wastewater. Department of Environmental Technology. Agricultural University of Wageningen.

Von Sperling M (1996). Comparison among the most frequently used systems for wastewater treatment in developing countries. *Wat. Sci. Tech.* **33** (3), 59-76.

Wang B (1991). Ecological waste treatment and utilization system: low-cost energy saving/generating, resources recoverable technology for pollution control in China. *Wat. Sci. Tech.* **24** (1), 8-19.

Yagoubi M, Foutlane A, Bourchich L, Jellal J, Wittland C and El Yachioui M (2000). Study on the performance of the wastewater stabilization pond of Boujaâd, Morocco. *J. Water Supply Res. Tech.* **49**(4), 203-209.

Zimmo OR, Al-Saed RM, Van der Steen P and Gijzen HJ (2002). Process performance assessment of algae and duckweed wastewater pond systems. *Wat. Sci. Tech.* **45**(1), 91-101.

Chapter Seven

Comparative Performance Studies of Macrophyte and Algal Ponds Using Medium Strength Sewage

Awuah E, Afealetse D-G, Appiah J, Lubberding HJ and Gijzen HJ

Comparative Performance Studies of Macrophyte and Algal Ponds Using Medium Strength Sewage

Comparative Performance Studies of Macrophyte and Algal Ponds Using Medium Strength Sewage

Abstract

The potential of water lettuce *(Pistia stratiotes)*, duckweed *(Spirodela polyrhiza)*, and algal (natural colonization) pond systems to treat domestic wastewater for the removal of faecal bacteria *(Escherichia coli*, coliforms, *Salmonella*, other enterobacteria and enterococci), organic matter and nutrient loads were evaluated for a period of 12 months in waste stablisation ponds operating in series of four for each treatment system. Biomass yield and protein content of macrophytes in the ponds were measured. When macrophytes and algae were visibly healthy, (performance was at its peak) an 8week intensive measurements of environmental conditions were determined at various depths (0.10m, 0.35m and 0.63m) in each pond of all 3-pond systems. Faecal bacteria were determined in the effluent from each pond in series. Neutral conditions and low DO concentrations were associated with the macrophyte ponds. The algal ponds had high DO concentrations. Removal of *E. coli, Salmonella* and other enterococci were the same in all three systems with no significant (p<0.5) differences during the one-year monitoring phase. Removal of all other faecal bacteria in the duckweed pond system was comparable to that of the algal pond system during this period. During the intensive measurements period however, both macrophyte pond systems performed better in the removal of coliforms and other enterobacteria than the algal ponds. BOD removal ranged from 83-99%, 88-97% and 79-83% in water lettuce, duckweed and algae respectively with the higher removal efficiencies occurring during the intensive measurement study. Total nitrogen removal was more than 80% in all pond systems. Total phosphorous removals were 42%, 62% and 61% in water lettuce, duckweed, and algal systems, respectively, during the one-year monitoring phase. Total biomass yield (wet weight) in the water lettuce and duckweed pond systems were respectively 503.6 and 286.6 tons /ha/year. Protein content in the macrophytes was highest in the second pond with water lettuce leaves having 22% and duckweed fronds having 34% protein.

Keywords: Water lettuce; duckweed; algae; faecal bacteria; performance

Introduction

Municipal and industrial wastewaters are largely discharged without treatment into surface waters throughout the developing world and examples of gross pollution are ubiquitous. The health implications on communities downstream cannot be overemphasized. According to the UN World Water Assessment Programme (UNESCO, 2003), 2 million people die from diarrhoea every year as a result of poor sanitation; most of them live in developing countries. Domestic wastewater treatment has thus become a matter of concern for governments of developing countries. In Africa where many people depend on surface water without treatment for potable water, quality criteria should be strict and wastewater discharges will have to be carefully controlled, particularly the removal of pathogenic organisms. With increasing urbanization in developing countries, municipalities must look for rational low-cost solutions to environmental quality problems if development goals are to be achieved without destroying natural resources. Such low cost facilities must, however, be able to provide high treatment efficiencies. Macrophyte-based waste stabilization ponds are being employed in wastewater treatment in many parts of the world as low cost technologies to promote resource recovery and wastewater re-use (Greenway, 2003). Wastewater characteristics and environmental conditions vary from place to place and the treatment efficiency of stabilization ponds may vary from country to country. Thus, assessment of treatment efficiencies of such technologies must be evaluated in a country before introduction on a large scale. Waste stabilization ponds, relying on natural processes based on bacteria and algae, for nutrients and pathogen removal have two vital objectives:

1. The elimination of pathogens associated with domestic wastewater, thus protecting public health and
2. The conversion of contaminants present in the wastewater into stable oxidized end products, which can be safely discharged into the environment without any adverse ecological effects.

Waste stabilization ponds are known to be low cost in construction, operation and maintenance and have been recommended for poor countries trying to improve on their sanitation (Von Sperling, 1996; Oakley *et al.,* 2000). Even though the operation and maintenance are low-cost, many of the ponds in Ghana are neglected and have broken down due to lack of incentives (Awuah *et al.,* 1996; Awuah *et al.,* 2002a). When macrophytes are used as part of waste stabilization ponds, the macrophytes and their associated biofilms play an important role in removing, transforming and storing nutrients. The plant cover reduces water velocity and turbulence causing filtration and settlement of particles and provides an increased surface area for attachment of microorganisms for degradation of organic materials. The high productivity and nutrient removal capability of macrophytes have created substantial interest in their use for wastewater treatment and resource recovery. Nutrients recovered in plant biomass can be used for aquaculture and other economic activities. Effluent, after this treatment, can further be used for irrigation without high algal suspended solids. This is an added advantage to wastewater treatment. However, the removal of pathogens is crucial for the safe reuse of effluents. The removal of pathogens, organic and nutrient loads in macrophyte ponds has not been extensively studied. The performance in comparison to that of algal ponds is unclear especially in terms of pathogen removal. Previous investigations using diluted sewage (low strength) showed that the removal of nutrients and organic matter were higher in the macrophyte ponds than in algal ponds, whereas faecal coliform removal was better in the algal ponds than in the macrophyte ponds (Awuah *et al.,* 2004a). The removal of enterococci, however, showed no significant differences between macrophyte and algal ponds (Awuah *et al.,* 2002b). The wastewater used in these studies was diluted and therefore represented low strength sewage. The aim of this study was to compare the overall performance between the macrophyte and algal pond systems using undiluted, medium strength sewage. More specifically, the objectives were to determine:

i. The environmental conditions in macrophyte and algal pond systems,
ii. The wastewater treatment efficiencies in macrophyte and algal pond systems in terms of faecal bacteria, nutrient and organic matter removal and
iii. The biomass yields and protein content of water lettuce and duckweed plants growing in the waste stabilization pond systems to measure their potential economic uses.

Materials and Methods

The study was conducted using a bench-scale continuous flow set up at the Kwame Nkrumah University of Science and Technology, (KNUST), Kumasi, Ghana in a tropical climate. The system consisted of 3 parallel treatment lines of 4 ponds each, involving the use of water lettuce (*Pistia stratiotes*), duckweed (*Spirodela polyrhiza*) and algae (natural colonization). A flow rate of $0.01m^3$/day was maintained in each treatment system. Each pond had a depth of 0.63m and surface area of $0.145m^2$ and hydraulic retention period of 7 days. Wastewater was collected at the influent grit chamber of the University's wastewater treatment plant at 7 GMT daily and put into the anaerobic pond, which fed into the 3 pond systems, by gravity in a continuous flow (Figure 1). Water lettuce ponds were maintained by harvesting macrophytes once in a week and duckweed ponds were harvested twice in a week.

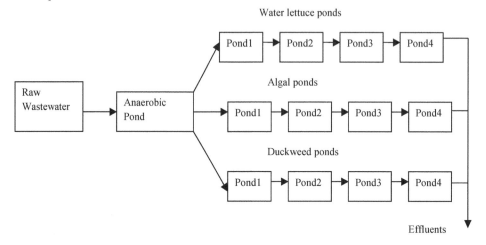

Figure 1 Schematic diagram of bench-scale macrophyte and algal waste stabilization ponds

Performance of the 3 systems was monitored during a 12 months period (low intensity sampling), with eight weeks intensive measurements. Parameters for the 12 month performance evaluation (low intensity) were: temperature, pH, dissolved oxygen (DO), total suspended solids (TSS), total dissolved solids (TDS), biochemical oxygen demand (BOD), chemical oxygen demand (COD), total phosphorous, ammonia, nitrate, nitrite and total organic and Kjeldahl nitrogen and indicators of faecal bacteria using *Escherichia coli*, coliforms, *Salmonella*, other enterobacteria and enterococci. These were measured monthly for one year in samples from the raw sewage, from the effluent of the anaerobic pond and the final effluents from each of the 3 pond systems between 7 and 8 (Greenwich Meridian time) GMT. After four months of algae colonization, when the performance was at its peak (plants were healthy and there was no disease or algae colonization in macrophyte ponds or the presence of floating algae in algal ponds), intensive performance measurements were made at each stage of treatment, for at least four times over a period of 8 weeks. Parameters measured during the intensive periods included TSS, turbidity, BOD, COD, ammonia, nitrate, nitrite, total Kjeldahl nitrogen, total phosphorous and faecal bacteria. Environmental conditions (temperature, pH, DO) were also measured at 8, 13 and 20 GMT at 0.1m (surface), 0.35m (middle) and 0.63m (bottom). Temperature and pH were measured with a portable electronic meter 322-B WTW. DO was measured with an oximeter, Oxi 323-B/set WTW. TDS and conductivity were measured with a conductivity meter/WTW. Organic matter and nutrient concentrations (BOD, COD, ammonia, nitrates, nitrites, total organic and Kjeldahl nitrogen, total phosphorous) were analysed according to Greenberg *et al.*, (1992).

Enumeration of *E. coli*, coliforms, *Salmonella* and other enterobacteria was done on chromocult agar after incubation at 37^0C for 24 hrs by spread plate method (Byamukama *et al.*, 2000). Enterococci were enumerated according to Niemi and Ahtiainen, (1995). The ponds were maintained by harvesting water lettuce once a week and duckweed twice a week. Biomass yield in the ponds was measured by weighing the amount of macrophytes harvested each week.

Protein content of the macrophytes was analysed by electrophoresis in the Department of Biochemistry, KNUST, Kumasi, Ghana.

Results

12 month (low intensity) performance studies

The temperature of the raw wastewater, of the anaerobic pond effluent and of the final effluent from all pond systems was similar (Table 1). The pH of the raw wastewater, the anaerobic pond effluent and the final effluents of the water lettuce and duckweed ponds were neutral, but the pH in the final effluent of the algal ponds was alkaline. The raw wastewater and the anaerobic pond were not completely anaerobic. The DO concentrations did increase along the series of ponds in all treatment systems, with the algal pond effluents having the highest DO levels. Total dissolved solids were high in both raw wastewater and in the anaerobic pond with a high variation. In all the pond systems, total dissolved solids declined along the series of ponds and were lowest in the water lettuce ponds (Table 1).

Table 1 Environmental conditions and characteristics of influent and effluents of macrophyte and algal ponds between 7-8GMT (Annual Means)

Parameter	Influent	Anaerobic pond	Water lettuce	Duckweed	Algae
Temperature (^{0}C)	27.4±3.4	26.9±3.8	27.1±3.7	27.0±3.7	27.4±3.5
pH	7.4±0.4	7.3±0.8	7.0±0.4	7.1±0.5	8.8±1.3
DO (mg/L)	1.3±1.3	2.0±1.9	3.5±1.5	3.5±1.1	4.4±3.6
TDS (mg/L)	1852±2202	2302±823	522±190	924±417	1076±530
SS (mg/L)	2270±160	97±49	52±36	24±17	93±53
COD (mg/L)	595±237	348±176	104±61	105±113	271±210
BOD (mg/L)	264±87	155±64	43±25	31±22	55±26
Total Kjeldahl N (mg/L)	78.9±32.5	90.5±32.1	4.6±3.5	3.8±3.7	8.4±5.7
Ammonia (NH_3-N) (mg/L)	59.0±25.2*	81.3±3.2	2.2±2.3	1.2±2.0	1.7±2.5
Organic-N (mg/L)	19.9±5.8	9.2±4.5	2.5±3.1	2.5±3.5	6.7±4.2
Nitrite (NO_2-N) (mg/L)	0.07±0.1	0.06±0.1	0.02±0.0	0.01±0.0	0.02±0.0
Nitrate (NO_3-N) (mg/L)	1.0±0.9	2.3±2.1	0.6±0.8	1.7±2.8	0.1±0.1
Total phosphorous (mg/L)	6.7±5.0	6.2±4.4	3.9±4.5	2.6±1.8	2.6±2.2

*± Standard deviation * High ammonia values obtained because the grit chamber was sometimes used for the discharge of septage, which has very high ammonia concentrations.*

The organic matter of the raw wastewater with an average BOD of 264mg/L (Table 1) was slightly higher than values reported by Metcalf and Eddy (1991) for medium strength sewage. High organic matter removal was observed in all three systems. Most of the time, the BOD and COD concentrations met the Environmental Protection Agency (Ghana) guidelines (EPA, 1995) of 50mg/L and 250mg/L respectively for BOD and COD (Figure 2). The organic matter removal efficiency in the water lettuce and duckweed ponds was higher (BOD and COD removal more than 80%) than in the algal ponds (BOD and COD removal of 79% and 54%, respectively). BOD, COD and TSS concentrations followed the same trend in the final effluents over the year for all pond systems (Figure 2).

Total Kjeldahl nitrogen and ammonia were reduced considerably (90% or more) in all pond systems. Nitrate and nitrite concentrations were negligible in all cases and well below the EPA guidelines. Total phosphorous concentrations in the influent and anaerobic pond were more than 6.0mg/L with a strong variation in all the samples analysed. Total phosphorous removal was 42% in the water lettuce, 62% in the duckweed and 61% in the algal pond systems (Table 1). The concentration of phosphorous in the effluent did not meet the Ghanaian EPA guidelines.

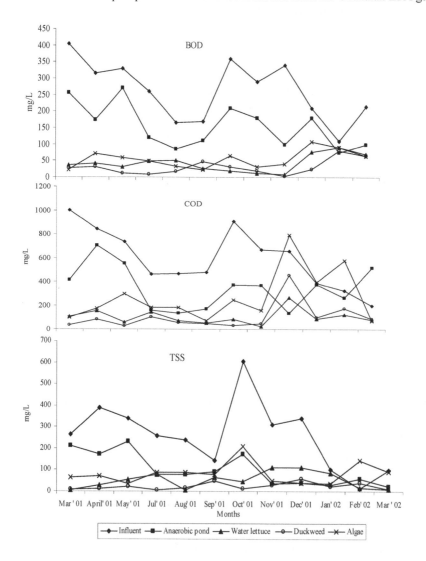

Figure 2 Removal of organic matter and TSS in macrophyte and algal ponds

Removal of *E. coli, Salmonella* and enterococci was not significantly (p<0.05) different among the 3 pond systems (Figure 4).

Water lettuce pond system had poor performance in the removal of other coliforms and enterobacteria. The removal of *E. coli*, other coliforms, other enterobacteria and enterococci in the duckweed pond system was comparable to that of algal pond system during the one-year monitoring (Figure 3).

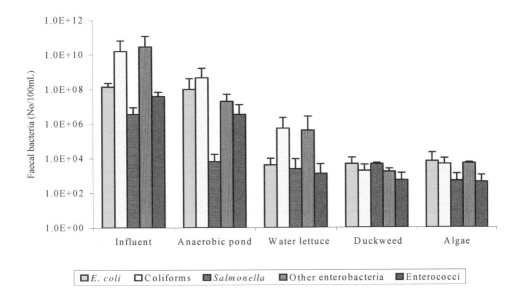

Figure 3 Annual means of faecal bacteria in influent and final effluent treatment pond systems

Eight week intensive performance studies

Environmental conditions

Table 2 Environmental conditions in the water lettuce pond system

		Time (in GMT)			Time (in GMT)		
		8	13	20	8	13	20
Pond	Depth	Temperature (^{0}C)			DO (mg/L)		
1	Surface (0.10m)	26.7±0.4	31.7±3.0	31.3±2.1	1.3±0.6	1.3±0.3	1.2±1.4
	Middle (0.35m)	26.5±0.7	29.5±1.8	31.0±2.0	1.4±0.5	0.9±0.4	1.1±1.3
	Bottom (0.63m)	26.2±0.4	29.3±1.5	30.4±2.2	1.3±0.4	0.8±0.5	1.4±1.1
2	Surface (0.10m)	26.8±0.4	31.5±2.7	30.9±2.0	2.2±1.3	1.4±0.4	1.3±1.2
	Middle (0.35m)	26.5±0.6	29.5±1.8	30.8±2.0	1.4±1.3	0.9±0.1	1.1±1.2
	Bottom (0.63m)	26.1±0.6	29.3±1.7	30.3±2.3	2.4±1.1	0.9±0.3	1.2±1.3
3	Surface (0.10m)	26.8±0.4	32.0±2.6	31.0±1.8	2.3±1.1	1.9±0.8	1.5±0.9
	Middle (0.35m)	26.6±0.6	29.7±2.0	30.7±1.9	1.4±1.1	1.0±0.3	1.2±1.0
	Bottom (0.63m)	26.3±0.6	28.7±1.7	30.1±2.1	2.5±0.8	0.9±0.3	1.2±1.2
4	Surface (0.10m)	26.9±0.4	32.2±2.9	30.6±1.9	2.5±0.6	2.9±1.5	2.1±0.4
	Middle (0.35m)	26.5±0.5	29.3±1.8	30.4±1.9	1.4±0.6	1.6±0.6	1.6±0.7
	Bottom (0.63m)	26.5±0.5	28.1±1.2	29.5±1.4	2.5±0.6	1.3±0.3	1.7±0.9
		pH			TDS (mg/L)		
1	Surface (0.10m)	7.2±0.1	7.1±0.1	7.2±0.0	467±488	227±227	714±413
	Middle (0.35m)	7.2±0.1	7.1±0.1	7.2±0.0	472±486	461±461	713±414
	Bottom (0.63m)	7.2±0.1	7.1±0.0	7.1±0.0	473±471	450±450	691±401
2	Surface (0.10m)	7.0±0.0	7.0±0.1	7.0±0.0	320±300	230±230	456±265
	Middle (0.35m)	7.0±0.0	7.0±0.1	7.0±0.0	318±299	380±380	457±266
	Bottom (0.63m)	7.0±0.1	7.0±0.1	7.0±0.0	317±295	379±379	448±256
3	Surface (0.10m)	6.7±0.0	6.8±0.1	6.8±0.0	235±188	278±278	291±171
	Middle (0.35m)	6.7±0.0	6.7±0.0	6.7±0.0	235±187	372±372	292±171
	Bottom (0.63m)	6.8±0.1	6.7±0.0	6.8±0.0	236±181	388±388	288±167
4	Surface (0.10m)	6.6±0.0	6.8±0.2	6.7±0.0	214±119	396±396	195±117
	Middle (0.35m)	6.6±0.0	6.6±0.0	6.7±0.0	215±119	457±457	195±117
	Bottom (0.63m)	6.5±0.1	6.6±0.0	6.6±0.0	215±118	454±454	196±118

Temperatures were generally higher on the surface than other depths for all the ponds in all 3-pond systems. The values ranged from 25-34^{0}C. There were no differences between water lettuce, duckweed and algal ponds (Table 2, 3 and 4). The macrophyte ponds had a neutral pH and low DO concentrations between 0.8 and 3.6mg/L. There were no diurnal fluctuations in pH and DO in the macrophyte ponds. In the water lettuce systems, the pH declined from pond 1 to pond 4 with slight differences towards acidic conditions between the first two ponds and the last two ponds (Table 2). The pH values did not change from pond 1 to pond 4 in the duckweed system (Table 3).

DO concentrations in the macrophyte ponds increased along the series of ponds with low values at the bottom. The pH and DO levels on the surface in the algal ponds increased along the series of ponds and showed diurnal fluctuations. Neutral pH and low DO levels were observed at the middle and at the bottom. Surface values in the algal ponds were different from bottom values. TDS in all three pond systems decreased along the series of ponds in the water lettuce ponds. The surface values were always lower than the middle and bottom portions of all pond systems. There was always an increase from 8GMT to 20GMT in all pond systems except in the last pond of the water lettuce system (Table 2, 3 and 4).

Table 3 Environmental conditions in the duckweed pond system

Pond	Depth	Time (in GMT)			Time (in GMT)		
		8	13	20	8	13	20
		Temperature (^{0}C)			DO (mg/L)		
1	Surface (0.10m)	27.6±0.7	34.0±2.4	30.1±1.4	2.2±1.5	1.1±0.7	0.8±0.8
	Middle (0.35m)	26.6±0.4	30.5±2.1	30.0±1.5	1.9±1.7	0.8±0.3	0.6±0.6
	Bottom (0.63m)	26.3±0.5	29.9±1.8	28.9±1.8	1.9±1.3	0.8±0.4	0.8±0.8
2	Surface (0.10m)	27.3±0.8	33.9±2.5	29.8±1.3	1.8±1.1	1.7±0.4	1.0±1.0
	Middle (0.35m)	26.4±0.2	30.2±2.0	29.8±1.4	2.0±1.0	0.8±0.4	0.7±0.7
	Bottom (0.63m)	26.1±0.4	29.7±1.8	28.9±2.1	2.2±1.0	0.7±0.5	0.7±0.7
3	Surface (0.10m)	27.3±1.0	33.7±2.6	29.9±1.3	2.3±0.7	2.2±0.3	1.8±0.4
	Middle (0.35m)	26.4±0.1	30.2±2.0	29.7±1.4	2.3±0.8	1.7±0.2	1.1±0.5
	Bottom (0.63m)	26.0 ± 10.3	29.6±1.7	29.4±2.1	2.5±0.6	1.9±0.4	1.1±0.5
4	Surface (0.10m)	27.0±0.6	33.2±3.1	29.7±1.3	2.4±0.7	3.6±1.2	2.6±0.2
	Middle (0.35m)	26.3±0.3	30.0±2.0	29.7±1.4	2.4±0.6	3.0±0.9	2.3±0.1
	Bottom (0.63m)	26.2±0.3	29.2±1.7	29.3±1.3	2.3±0.5	2.5±0.3	1.7±1.1
		pH			TDS (mg/L)		
1	Surface (0.10m)	7.2±0.0	7.3±0.1	7.2±0.0	220±455	220±46	746±449
	Middle (0.35m)	7.2±0.0	7.2±0.1	7.2±0.0	478±466	478±385	746±449
	Bottom (0.63m)	7.1±0.0	6.9±0.2	7.0±0.1	475±467	475±377	725±440
2	Surface (0.10m)	7.3±0.0	7.3±0.0	7.3±0.0	356±304	356±175	538±449
	Middle (0.35m)	7.2±0.0	7.3±0.0	7.3±0.0	351±305	351±230	538±449
	Bottom (0.63m)	7.2±0.0	7.2±0.0	7.2±0.0	348±303	348±229	531±440
3	Surface (0.10m)	6.9±0.0	7.0±0.1	7.0±0.1	119±223	119±338	410±255
	Middle (0.35m)	6.9±0.0	6.9±0.1	6.9±0.1	267±223	267±286	411±256
	Bottom (0.63m)	6.9±0.1	6.9±0.0	6.9±0.0	264±225	264±313	406±254
4	Surface (0.10m)	7.0±0.1	7.2±0.1	7.0±0.1	197±205	197±589	368±227
	Middle (0.35m)	7.0±0.1	6.9±0.2	7.0±0.1	242±208	242±528	367±227
	Bottom (0.63m)	6.9±0.1	6.7±0.1	7.6±0.9	239±209	239±526	363±225

Table 4 Environmental conditions in the algal pond system

Pond	Depth	Time (in GMT) 8	13	20	Time (in GMT) 8	13	20
		Temperature (^{0}C)			**DO (mg/L)**		
1	Surface (0.10m)	27.0±1.1	34.1±2.9	30.0±2.0	2.1±0.7	6.6±4.0	1.1±1.3
	Middle (0.35m)	26.1±0.3	30.1±2.4	29.8±2.2	2.3±0.7	2.3±1.0	1.1±1.3
	Bottom (0.63m)	26.1±0.5	29.5±2.4	29.5±2.2	2.5±0.7	1.9±1.5	1.1±1.3
2	Surface (0.10m)	26.9±0.9	33.8±0.1	30.0±2.0	2.6±0.6	6.9±4.7	1.4±1.1
	Middle (0.35m)	26.1±0.3	30.1±0.1	29.7±2.0	2.3±0.8	2.6±1.1	1.2±1.3
	Bottom (0.63m)	26.0±0.3	29.4±0.0	29.7±2.1	2.4±0.5	1.9±0.9	1.2±1.4
3	Surface (0.10m)	26.6±0.6	33.7±3.4	29.9±1.9	2.3±0.5	5.7±3.1	2.5±1.1
	Middle (0.35m)	26.1±0.4	29.5±2.2	29.6±1.9	2.3±0.5	1.8±0.7	2.2±0.7
	Bottom (0.63m)	25.8±0.4	28.7±1.9	29.2±1.9	2.4±0.5	1.7±1.2	1.4±1.5
4	Surface (0.10m)	26.5±0.4	33.5±3.4	30.0±1.9	2.5±0.4	5.6±3.3	3.9±3.3
	Middle (0.35m)	26.1±0.3	29.6±2.2	30.0±2.0	2.6±0.4	2.8±1.6	2.9±2.6
	Bottom (0.63m)	26.0±0.4	28.6±1.6	29.3±1.9	2.7±0.4	1.7±0.5	1.3±1.4
		pH			**TDS (mg/L)**		
1	Surface (0.10m)	7.7±0.1	8.3±0.7	7.7±0.1	168±420	168±17	574±323
	Middle (0.35m)	7.6±0.1	7.6±0.1	7.7±0.2	388±421	388±438	574±322
	Bottom (0.63m)	7.2±0.2	7.2±0.1	7.3±0.3	382±426	382±434	553±301
2	Surface (0.10m)	8.0±0.3	9.5±0.7	8.2±0.0	110±247	110±6	381±231
	Middle (0.35m)	7.8±0.1	7.9±0.3	8.2±0.0	266±247	266±335	390±230
	Bottom (0.63m)	7.3±0.3	7.6±0.3	7.8±0.3	261±239	261±333	405±239
3	Surface (0.10m)	8.8±0.4	10.4±0.1	9.2±0.4	111±191	111±3	329±231
	Middle (0.35m)	8.5±0.1	8.7±0.3	9.1±0.4	230±193	230±259	329±230
	Bottom (0.63m)	7.9±0.1	8.3±0.1	8.4±0.2	335±217	335±257	328±239
4	Surface (0.10m)	9.1±0.4	10.2±0.3	9.6±0.4	256±192	256±3	339±204
	Middle (0.35m)	9.0±0.3	9.2±0.3	9.4±0.4	335±192	335±238	334±206
	Bottom (0.63m)	8.8±0.4	9.0±0.3	9.1±0.2	328±192	328±236	327±202

±: *Standard deviation*

Organic and nutrient load removal

Organic and nitrogen loads decreased substantially from pond 1 to pond 4. Organic matter was efficiently removed in water lettuce and duckweed ponds (>95%). For the water lettuce, the effluent quality in the first pond - in terms of BOD and COD - met Ghanaian EPA guideline values of 50 and 250mg/L, respectively. For duckweed, the second pond effluent quality was enough to meet these guidelines values. A similar tendency was observed in TDS and turbidity. In all cases, the Ghanaian EPA guidelines were met except for algal ponds. Total nitrogen and ammonia decreased dramatically from pond 1 to pond 4 in all pond systems (>90% reduction). Nitrite and nitrate concentrations increased along the series of ponds for the macrophyte systems. There was no difference in all four algal ponds (Table 5).

Total phosphorous did not decrease much along the series of ponds. Turbidity reduced drastically and correlated well with the suspended solids removal in the macrophyte systems. Turbidity and TSS levels in the algal ponds were, however, higher than those in the macrophyte ponds (Table 5).

Table 5 Intensive performance measurements of macrophyte and algal waste stabilization ponds

Treatment system	BOD (mg/L)	COD (mg/L)	TSS (mg/L)	NH3-N (mg/L)	NO$_2$-N (mg/L)	NO$_3$-N (mg/L)	Total N (mg/L)	Total P (mg/L)	Turbidity (NTU)
Anaerobic pond	343±140	683±30	1181±404	108.5±7.0	0.03±0.0	0.05±0.0	163±12.7	4.1±0.4	1115±5
Water lettuce									
Pond 1	49.3±2.3	119±32	424±222	63.4±1.7	0.01±0.0	0.04±0.0	66.9±1.2	4.8±0.8	87.1±22.9
Pond 2	29.3±3.1	72±4	105±48	28.2±0.4	0.01±0.0	0.23±0.1	32.3±0.5	2.8±0.5	12.6±4.2
Pond 3	5.3±1.5	39±5	77±2	14.7±0.9	0.66±0.3	0.63±0.0	17.1±1.1	5.1±0.1	4.8±2.4
Pond 4	3.3±0.6	17±0	9±5	7.0±0.3	0.22±0.2	0.54±0.0	9.2±0.4	2.8±0.2	7.9±0.6
Duckweed									
Pond 1	76±4.0	144±33	135±5	98.1±4.9	0.04±0.0	0.02±0.0	102.3±4.8	4.0±0.1	40.6±7.5
Pond 2	31±2.3	55±14	41±19	41.9±1.3	0.03±0.0	0.13±0.0	44.5±1.7	4.2±0.4	3.3±0.9
Pond 3	14±2.5	36±4	12±4	19.2±0.5	0.56±0.1	0.26±0.0	21.3±0.5	4.2±0.4	3.1±0.2
Pond 4	11±1.0	25±0	12±10	7.8±0.4	0.03±0.0	0.40±0.0	9.3±0.7	4.0±0.1	2.1±0.1
Algae									
Pond 1	100±12.0	231±19	85±9	46.7±1.8	0.01±0.0	0.16±0.0	51.5±2.3	4.2±0.4	58.3±27.1
Pond 2	76±2.0	164±13	17±6	17.9±1.1	0.02±0.0	0.02±0.0	22.0±2.6	5.2±0.1	17.5±1.9
Pond 3	61±3.5	146±9	151±3	3.8±0.8	0.02±0.0	0.18±0.1	8.1±1.0	4.8±2.5	39.9±12.7
Pond 4	55±5.6	132±12	114±24	0.3±0.4	0.04±0.0	0.06±0.0	5.5±0.8	4.2±0.3	34.5±5.8
EPA (Ghana) guidelines	**50**	**250**	**50**	**1.5**	-	**10**	-	**2**	**75**

The number of faecal bacteria declined along the series of ponds. The general trend of faecal bacteria removal along the series of ponds was not very different among the different pond systems from pond 1 to pond 3. Most of the elimination of faecal bacteria occurred in the last pond. The performance in the removal of faecal bacteria in the macrophyte pond systems was comparable to that algal pond system. Removal of $E.$ $coli$ in terms of $\log_{10}$ was the same for the algal and water lettuce pond systems. Other coliforms were almost eliminated in the water lettuce ponds. The removal ($\log_{10}$) of other coliforms in the algal ponds was the same as that of the duckweed ponds. Removal ($\log_{10}$) of other enterobacteria in the macrophytes ponds was better than that in the algal ponds (Table 6-8).

Table 6 Average faecal bacteria numbers in the effluent of water lettuce ponds

Bacteria (No/100mL)	Anaerobic pond	Pond 1	Pond 2	Pond 3	Pond 4	$\log_{10}$ removal
E. coli	$3.7x10^6$	$7.7x10^6$	$2.6x10^4$	$5.1 x10^4$	11	5
Coliforms	$6.2x10^6$	$5.9x10^6$	$6.7x10^4$	$3.9x10^5$	1.0	6
Salmonella	$1.2x10^6$	$4.2x10^6$	$5.3x10^4$	$1.0x10^5$	$2.0x10^2$	4
Other enterobacteria	$1.3x10^7$	$2.4x10^8$	$1.2x10^7$	$4.7x10^6$	1.0	7

Table 7 Faecal bacteria numbers in the effluent of duckweed ponds

Bacteria (No/100mL)	Anaerobic pond	Pond 1	Pond 2	Pond 3	Pond 4	$\log_{10}$ removal
E. coli	$3.7x10^6$	$1.9x10^6$	$6.7x10^5$	$6.8x10^4$	$1.9x10^2$	4
Coliforms	$6.2x10^6$	$2.0x10^6$	$1.9x10^5$	$8.3x10^4$	$5.0x10$	5
Salmonella	$1.2x10^6$	$6.6x10^7$	$4.0x10^3$	$7.5x10^4$	8.0	6
Other enterobacteria	$1.3x10^7$	$8.7x10^7$	$1.3 x10^6$	$2.8x10^4$	$1.7x10^2$	5

Table 8 Average faecal bacteria numbers in the effluent of algal ponds

Bacteria (No/100mL)	Anaerobic pond	Pond 1	Pond 2	Pond 3	Pond 4	$\log_{10}$ removal
E. coli	$3.7x10^6$	$1.7x10^7$	$1.1x10^6$	$6.1x10^4$	10.0	5
Coliforms	$2.4x10^6$	$1.0x10^7$	$7.6x10^6$	$2.7x10^6$	10.0	5
Salmonella	$1.2x10^6$	$5.5x10^6$	$3.8x10^6$	$1.0x10^5$	$5.3x10^2$	4
Other enterobacteria	$1.3x10^7$	$2.0x10^7$	$9.7x10^8$	$2.8x10^6$	$1.2x10^4$	3

Biomass yield in macrophyte ponds

Growth of macrophytes was initially low in March and progressively increased to August. The growth in March was the lowest (Figure 5). The macrophytes were probably adjusting to the new environment. The growth yield of water lettuce decreased from high in the first two ponds to low values in the last ponds (Figure 5). Duckweed yields were high and similar in the 2nd and 3rd ponds; the lowest yield was observed in the first pond (Figure 5). The yield of water lettuce was higher than that of duckweeds.

The protein content of the macrophytes decreased along the series after pond 2. Pond 2 had the highest protein content for both duckweed fronds and water lettuce leaves, with values of 34% and 22%, respectively. At the same stage of treatment in the series of ponds, duckweed fronds had higher protein content than water lettuce leaves (Table 9).

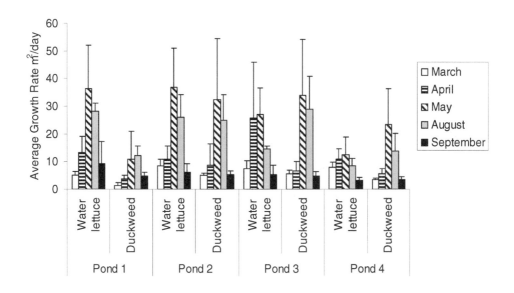

Figure 5 Daily biomass (grams) yields of water lettuce and duckweed ponds

Table 9 Total biomass yields and protein content of water lettuce and duckweed leaves

Pond System	Yield (tonnes fresh wt.ha^{-1} yr^{-1})	Protein content (%)
Water lettuce		
Pond 1	153.3	20
Pond 2	157.0	22
Pond 3	113.2	17
Pond 4	80.1	8
Total	**503.6**	
Duckweed		
Pond 1	45.6	28
Pond 2	87.6	34
Pond 3	84.0	29
Pond 4	69.4	15
Total	**286.6**	

Discussion

Environmental conditions

The environmental conditions reported here are similar to those in the same pond systems using low strength sewage (Awuah *et al.*, 2004a), except that the pH in the last two water lettuce ponds was neutral instead of acidic conditions. TDS was also higher in the medium strength sewage study than in the low strength sewage study (Table 2). DO levels were much higher in the previous low strength sewage study than in the present study. The prolific growth, due to increased nutrient levels, may have contributed to high oxygen levels consumption in the water lettuce, duckweed and algal ponds. Awuah *et al.*, (2004a) have explained the alkaline conditions observed in the algal ponds in detail and attributed the low pH in water lettuce ponds to under utilization of CO_2 for photosynthesis in the wastewater. The neutral conditions thus observed could be proper utilization of CO_2 produced in the wastewater.

The oxygen level in the anaerobic pond was 2.3±1.9 mg/L, which means it was not anaerobic at all. This was due to rapid mixing and atmospheric diffusion during the pouring of wastewater from the grit chamber into the pond.

Organic matter removal

The removal of organic matter was always better in the macrophyte ponds than in the algal ponds. The organic matter removal in the macrophyte ponds was so good that the effluent from the 1st water lettuce pond and the 2nd duckweed pond already met the Ghanaian EPA guidelines. This could reduce the retention period to 7 and 14 days, respectively. However, other parameters did not meet the recommended guidelines at this stage of treatment. The algal pond could meet the same criterion only in the last pond. Organic matter removal could be due to several physical and biochemical processes, including sedimentation of suspended organic matter in the wastewater (Gannon et al., 1983; Iqbal 1999), the oxidation of organic matter by heterotrophic bacteria in wastewater (Iqbal 1999; Maynard et al., 1999) and anaerobic degradation (Bonomo et al., 1997) in the bottom sediments of the ponds (Table 2, 3 and 4). The macrophyte cover provided quiescent conditions to reduce turbulence and enhanced sedimentation of suspended materials. In addition, the root zone provided a place of attachment for heterotrophic bacteria to degrade organic materials attached to the roots. The algae also provided surface area for attachment and oxygen for the breakdown of organic matter and for nitrification. Some of the algae also settled to the bottom. Since algae were not harvested the remaining algae in the wastewater contributed to high suspended solids, organic matter and turbidity (Table 1 and 5).

Nitrogen removal

A high removal of nitrogen, mainly present as ammonia and organic nitrogen was observed in all pond systems. In the anaerobic pond the concentration of organic nitrogen decreased, while the ammonia-nitrogen concentration increased (Table 1), suggesting that organic nitrogen was converted into ammonia. The ammonia concentrations gradually decreased from pond 1 to pond 4. The concentrations of nitrites and nitrates in all the ponds were low. Since anaerobic conditions were observed in the sediments, nitrates and nitrites could have been lost as N_2 through denitrification. The macrophytes could have also rapidly assimilated nitrites and nitrates when produced. Another reason could be due to the direct uptake of ammonia by both macrophytes and algae. Some studies on nitrogen removal in waste stabilization ponds have reported that about 30% of the total nitrogen load was assimilated by duckweed (Zimmo, 2003). In the algal ponds, where the removal of ammonia was higher than in the macrophyte ponds, ammonia volatilisation could be another mechanism, especially at high pH. The highest pH (10) in the algal ponds was observed in the last pond. Denitrification in addition to plant uptake and volatilisation can also occur in the all pond systems and reduce the concentration of nitrogen in wastewater treatment systems.

Phosphorous removal

The phosphorous concentrations in the final effluents did not meet the Ghanaian EPA guidelines. The low removal of phosphorus in these studies (Table 5) showed that macrophytes used for wastewater treatment may not be able to adequately remove total phosphorous.

In the aquatic environment, phosphorous is normally removed by the following mechanisms: plant uptake, adsorption by clay particles and organic matter, chemical precipitation with Ca^{2+}, Fe^{3+}, Al^{3+} and microbial uptake (Iqbal, 1999). Heavy metal concentrations in the wastewater used were generally low (Awuah et al., 2002a) and therefore could not be a reason for the phosphorous removal. The removal of total phosphorous observed in this study is likely due to settling of phosphorous associated with solids present and direct uptake. Kansiime and Nalubega (1999) who made an estimation of phosphate uptake by macrophytes in Nakivubo swamp in Uganda reported that phosphorous removal in papyrus wetlands was due to macrophyte uptake and subsequent harvesting. Ultimate P-removal is only possible by plant harvesting (as was the case in duckweed and water lettuce ponds). Algae are also sequesters of phosphorous as phosphates, in fresh water systems (Gaiser et al., 2004) and phosphorous could be removed by sedimentation in algal ponds. Mara and Pearson (1986) reported that sedimentation of organic phosphorous, as macrophyte and algal biomass, was an important mechanism for phosphorous removal. However, during decay, phosphorous may be released back into the water medium as phosphate. The low removal of phosphorous observed in algal ponds may be due to re-suspension of phosphorus in the algal biomass at high pH and DO (Mara and Pearson, 1986).

The removal of faecal bacteria and their profiles in macrophyte and algal ponds

The removal of faecal coliforms using medium strength sewage is contrary to what we reported in low strength sewage (Awuah et al., 2004a), where the removal of faecal coliforms in algal ponds was significantly different from the removal in macrophyte ponds. The increase in faecal bacteria numbers recorded on chromocult agar could be due to its ability to activate damaged bacteria (Byamukama et al., 2000). The chromocult agar is able to differentiate four different groups of faecal bacteria and it holds a promising future for faecal bacteria enumeration. Its use may also result in reported cases of higher faecal bacteria populations than in earlier works (Metcalf and Eddy, 1991; Awuah et al., 1996; 2003; 2004a). The removal of faecal bacteria in the macrophyte ponds was comparable to that of algal ponds in terms of E. coli, other coliforms and enterococci in the one-year monitoring phase. Earlier comparative performance studies using enterococci as indicators also showed that there were no differences between macrophyte and algal ponds (Awuah et al., 2001; 2002b).

During the intensive measurements study, faecal coliform removal in the first three ponds was almost the same. It appeared that faecal bacteria removal was dependent on retention time. The environmental conditions in the macrophyte ponds showed the absence of turbulence recorded in low turbidity readings (Table 8), which support the occurrence of sedimentation. Earlier work by Awuah et al., (2004a) in a low strength sewage performance study showed that >99% of the faecal coliforms were found in the sediment. In the algal ponds, the alkaline pH (8-10) and the fluctuating pH (from 7 to 9) (Table 4) might have contributed to the removal of faecal bacteria. E.coli is susceptible to high pH (10-11) and fluctuating pH (7-9) (Awuah et al., 2003: Parhad and Rao, 1974: Awuah et al., 2004). Macrophytes and algae may also provide surface areas for attachment and in that way contribute to faecal bacteria removal.). In the algal ponds high amounts of faecal bacteria could get attached to algae and attachment of faecal bacteria to algae could be important. This is because these faecal bacteria would then be close to the site of generation of harsh environmental conditions such as high pH and DO for the maximum impact to be felt.

In papyrus wetland, which is similar to macrophyte ponds, sedimentation (Kansiime and van Bruggen, 2001) attachment to surfaces (Spira et al., 1981), filtration (Gersberg, et al., 1987), predation (Stott *et al.,* 2001) and natural die-off (Oron *et al.,* 1987) have been shown to be responsible for faecal bacteria removal.

Growth yields in water lettuce and duckweed ponds

The growth yield of the macrophytes used in this study is several times higher than that of other agricultural crops (Oron *et al.,* 1987; Reddy and Debusk, 1984). The growth yields obtained in our study (Figure 5) is about ten times higher than maize growth yield per hectare per year in the tropics (FSAU/SCF, 2000). The high productivity in the water lettuce ponds is due to the high nutrient levels present in the wastewater. The high growth yields in duckweeds may be attributed by high nutrients present in the wastewater and their ability to absorb nutrients by their fronds, which are constantly in contact with the wastewater (Oron *et al.,* 1987). High productivity of these macrophytes is essential for commercial purposes. Water lettuce is used as a fodder for pigs and as herbal preparations for asthmatic patients and for cure of skin diseases in Ghana. Duckweed has no current use in Ghana, but some chickens have been seen eating the fronds in Ghana. However, the chickens preferred worms and maize to the duckweed fronds (Mr. Ofori Donkor personal communication). The duckweed alone therefore cannot be used as poultry feed. It must be mixed with other sources of feed like maize and dried fish. Gijzen and Khonker, (1997) reported up to 45% protein in duckweed and Sharma and Sridhar (1985) reported of 25% protein in water lettuce. The protein content in the water lettuce leaves and duckweed fronds in our study were maximally 22% and 34%, respectively and comparable to those in groundnuts (30%) and peas (23%) (Breeman, 1998). The protein content in the duckweed can thus be harnessed and used for poultry as it is been done in Zimbabwe (IWSD News, 1999). Fish farming is practiced in many parts of the country and duckweed can be used as a feed for fish as practiced in Bangladesh (Gijzen and Khonker, 1997).

Conclusion

The Macrophyte-based waste stabilization ponds can be used in the treatment of domestic wastewater with a high efficiency of organic and nitrogen load removal if the pond systems are well managed. They could be equally capable of removing pathogens as in algal ponds contrarily to our findings in low sewage strength studies. The removal of phosphorous is poor in both macrophyte and algal ponds and may require policy statements and legislation to reduce phosphorous levels in domestic wastewater from source. Conditions created in macrophyte ponds in this study were not too harsh to cause faecal bacteria die-off. The removal of faecal coliforms in the macrophyte ponds could be due to sedimentation, natural and die-off due to the long retention periods. In the algal ponds extreme environmental conditions of high pH and DO levels could be the major cause of faecal bacteria removal.

Acknowledgement

This research was made possible by grants from the Netherlands government through the SAIL Foundation.

References

Awuah E, Nkrumah E and Monney JG (1996). Performance of the Asokwa waste stabilization ponds and the conditions of other sewage treatment plants in Ghana. *J. Univ. Sci. Tech.* **16**(1/2), 121-126.

Awuah E, Asante K, Anohene F, Lubberding HJ and Gijzen HJ (2001) Environmental conditions in macrophyte and algal-based domestic wastewater pond systems. *Wat. Sci. Tech.* **44**(6), 11-18.

Awuah E, Kuffour AR, Lubberding HJ and Gijzen HJ (2002a). Characterization and management of domestic wastewater in two suburbs of Kumasi. In: *Proceedings of Water and Health Conference*, Ottawa, Canada, September 2002. pp 367-384.

Awuah E, Asante K, Lubberding HJ and Gijzen HJ (2002b). The effect of pH on enterococci removal in *Pistia,* duckweed and algal-based stabilization ponds for domestic wastewater treatment. *Wat. Sci. Tech.* **45**(1), 67-74.

Awuah E, Boateng J, Lubberding HJ and Gijzen HJ (2003). Physico-chemical parameters and their effects on pathogens in domestic wastewater In: *SERR 2 Proceedings.* Elmina, September, 2002. Dzisi KA and Fiabge YAK eds. September, 2003, School of Engineering. KNUST Kumasi, Ghana pp 113-123

Awuah E, Oppong-Peprah M, Lubberding HJ and Gijzen HJ (2004a). Comparative performance studies of macrophyte and algal-based stabilization ponds. *J. Toxicol. Environ. Health.* Part A, **67**, 1-13.

Awuah E, Lubberding HJ and Gijzen HJ (2004b). The effect of stable pH and pH fluctuations on faecal bacteria in domestic wastewater. In: *SERR 3 Proceedings.* - Dzisi KA and Fiabge YAK eds. September, 2003, School of Engineering. KNUST Kumasi, Ghana pp 141-149

Bonomo L, Pastorelli G and Zambon N (1997). Advantages and limitations of duckweed-based wastewater pond systems. *Wat. Sci. Tech.* **35**(5), 239-246,

Breeman S (1998). Unroasted peanuts a secrete weapon? Internet site: http://www.alpet.co.za/allpgeons/steven/peanuts.htm.

Byamukama D, Kansiime F, Mach RL and Farnleitner H (2000). Determination of *Escherichia coli* contamination with chromocult coliform agar showed a high level of discrimination efficiency for differing faecal pollution levels in tropical waters of Kampala, Uganda. *Appl. Environ. Microbiol.* **66**, 864-868.

EPA Ghana (1995). Effluent guidelines for industrial waste and sewage treatment. *Daily Graphic* March 1995. 1p.

FSAU/SCF (2000) Food economy baseline profiles lower Juba Agro Pastoral: Maize and cattle Internet site: http:// www.unsomalia.net/fsau/repoerts/Bprofile_Lower JubaA-P.pdf

Gaiser EE, Scinto DJ, Richards JA, Jayachandran K, Childers DL, Trexler X and Jones RD (2004) Phosphorous periphyton mats provides the best metric detecting low-level P enrichment in and oligotrophic wetland. *Wat. Res.* **38**, 507-516.

Gannon JS, Buse K and Schillinger J (1983). Faecal coliform disappearance in a river impoundment. *Wat. Res.* **17**, 1595-1601.

Gersberg RM, Lyon SR, Brenner R and Elkins BV (1987). Fate of viruses in artificial wetlands. *Appl. Environ. Microbiol.* **83**, 731-736.

Gijzen HJ and Khonker M (1997). An overview of the ecology, physiology cultivation and application of duckweed. Literature review. Report No. 0896. Duckweed research project. Bangladesh, pp. 69.

Greenberg AE, Clesceri LS and Eaton AD (1992). Standard methods for examination of water and wastewater. 18th edition, American Public Health Association. American Water Works Association. Water Environmental Fed. Washington. D.C.

Greenway M (2003). Suitability of macrophytes for nutrient removal from surface flow constructed wetlands receiving secondary treated sewage effluent in Queensland, Australia. *Wat. Sci. Tech.* **48**(2), 121-128.

Iqbal S (1999). Duckweed aquaculture. Potentials possibilities and limitations for combined wastewater treatment and animal feed production in developing countries. SANDEC report no. 6/99. 1-91.

IWSD News (1999). An investigation into the Potential use of duckweed based waste stabilization ponds for wastewater treatment in small centres in Zimbabwe. Issue No. 6. December, 5p.

Kansiime F and van Bruggen JJA (2001). Distribution and retention of faecal coliforms in the Nakivubo wetland in Kampala. *Wat. Sci. Tech.* **44** (6), 199-206.

Kansiime F and Nabulega M (1999). Wastewater treatment by natural wetland: the Nakivubo swamp, Uganda-Process and implications, PhD dissertation, UNESCO-IHE Delft, A.A. Balkema Publishers, The Netherlands, 300p.

Mara DD and Pearson HW (1986). Artificial fresh water environment; Waste stabilization ponds. In: Rehm HJ and Reed G (eds), *Biotechnology-A comprehensive Treatise*, **Vol.8.** pp. 177-206.

Maynard HE, Ouki SK and Williams SC (1999). Tertiary Lagoons: A review of removal mechanisms and performance. *Wat. Res.* **33**, 1-13.

Metcalf and Eddy (2003). Wastewater engineering: Treatment, disposal and reuse. 3rd Edition. McGraw-Hill, New York, 1334p.

Niemi RM and Ahtiainen. J (1995). Enumeration of intestinal enterococci and interfering organisms with Slanetz-Bartley agar, KF streptococcus agar and the MUST method. *Lett. Appl. Microbiol.* **20**(2), 92-7.

Oakley SM, Docasangre, Flores C, Monge J and Estrada M (2000). Waste stabilization ponds in Central America. The experience of El Salvador, Guatemala, Honduras and Nicaragua. *Wat. Sci Tech.* **42**(1), 51-58.

Oron G, deVegt A and Porath D (1987). The role of the operation regime in wastewater treatment with duckweed. *Wat. Sci. Tech.* **19**, (1) 97-105.

Parhad NM and Rao NU (1974). Effect of pH on survival of *E. coli. Wat. Pollut. Contr. Fed.* **46** (5), 980-986.

Reddy KR and DeBusk WF (1984). Growth characteristics of aquatic macrophytes cultured in nutrient enriched water. Water hyacinth, water lettuce and pennyworth. *Economic Botany*, **38**, 229 - 239.

Sharma BM and Sridhar MKC (1985). Some observations on oxygen changes in lake covered with *Pistia stratiotes. Wat. Res.* **19**, 935-939.

Spira WM, Hug A, Ahmed QS and Saeed YA (1981). Uptake of *Vibrio cholerae* Biotype eltor from contaminated water by water hyacinth (*Eichhornia crassipes*). *Appl. Environ. Microbiol.* **42**, 50-553.

Stott R, May E, Matsushita E and Warren A (2001). Protozoa predation as a mechanism for the removal of *Cryptosporidium* oocysts from wastewaters in constructed wetlands. *Wat. Sci. Tech.* **44**(6), 191-198.

UNESCO (2003). World water assessment program. -UN water development report. pp. 102-125.

Von Sperling M (1996). Comparison among the most frequently used systems for wastewater treatment in developing countries. *Wat. Sci. Tech.* **33** (7), 59 - 76.

WHO (1989). Health guidelines for the use of wastewater in agriculture and aquaculture.

Technical report series. 778 World Health Organisation, Geneva.

Zimmo O (2003). Nitrogen transformations and removal mechanisms in algal and duckweed waste stabilization ponds. PhD Thesis dissertation, UNESCO-IHE Delft, A.A. Balkema Publishers, The Netherlands, 133p.

Chapter Eight

The Role of Attachment in the Removal of Faecal Bacteria from Macrophyte and Algal Waste Stabilization Ponds

Awuah E, Macauley M, Pinkrah R, Lubberding HJ and Gijzen HJ
The Role of Attachment in the Removal of Faecal Bacteria from Macrophyte and Algal Waste Stabilization Ponds

The Role of Attachment in the Removal of Faecal Bacteria from Macrophyte and Algal Waste Stabilization Ponds

Abstract

Attachment of bacteria occurs readily on most available surfaces. The question addressed in this study is whether this mechanism plays a role in pathogen removal in macrophyte and algal waste stabilization ponds. An attempt was made to answer this question in trials on a batch-scale, a bench-scale continuous flow system in Ghana and by using a pilot-scale continuous flow system in Colombia, South America. The results showed that faecal bacteria attach to walls of containers holding wastewater, water lettuce roots and leaves, duckweed fronds and algae. When the die-off rates and mass balance of faecal bacteria on various surfaces in batch-scale incubations were studied, die-off was observed immediately after attachment. Higher die-off was observed in surfaces in the algal ponds. Most of the viable bacteria were found attached to water lettuce roots and to suspended algae. Harvesting of macrophytes removed <1% of viable faecal bacteria in continuous flow ponds in Ghana and in Colombia. In comparison to percentage of faecal bacteria attachment to surfaces with total viable bacteria numbers, attachment was substantially found to contribute to faecal bacteria removal. Attachment and subsequent settling of suspended solids contributes to prolonged retention of faecal bacteria in stabilization ponds, and as such provides the conditions for die-off.

Key words: Attachment; water lettuce; duckweed; algae; faecal bacteria; wastewater

Introduction

Throughout Africa, the chronic shortage of water has had grave consequences for the inhabitants of the continent. Despite the commendable activities of many governmental and non-governmental agencies, millions of people continue to suffer as the meagre water supplies dwindle and become increasingly polluted. The situation in Ghana as in most developing countries is quite serious since the country is faced with a rapidly growing population accompanied by an increased demand for water supply and sanitation facilities. This has heightened the need to optimise the utilisation of available water resources including the treatment and re-use of wastewater. The current national water supply and sanitation coverage figures show that only 41% of the people have access to potable water and 29% to sanitation (Mime Consult, 2004). This has resulted in untreated domestic wastewater being dumped into the environment untreated. In addition, the end of pipe treatment systems currently in place, which are considered safe sanitation systems, may not produce the desired results to safeguard the environment and public health. Awuah *et al.*, (2002) reported that almost all the wastewater treatment plants in Ghana have broken down and farmers use the raw wastewater for vegetable crop production without treatment. Some farmers also use waste stabilization ponds, which are not functioning properly, as fishponds. These practices pose health risks to consumers. If the wastewater is to be re-used then it must be treated. The use of macrophytes in wastewater treatment is gaining recognition on the continent because it allows for resource recovery from the use of macrophyte as animal feed and for aquaculture (Nhapi *et al.*, 2003). This practice could provide an incentive for operating domestic wastewater treatment systems. However, if these systems are going to be used in developing countries then their pathogen removal efficiencies should be optimised to reduce enteric diseases, which are plaguing the continent. In addition, if the macrophytes are to be used for animal feed and aquaculture, then the amount of pathogens attached to the macrophytes should be known and preventive measures put in place for health reasons.

The extent of pathogens attachment to macrophytes and to algae and the contribution to the removal of pathogens in waste stabilization ponds is not well known. The contribution of this mechanism to pathogen removal needs to be investigated for improvement in engineering designs and operation and maintenance practices of macrophyte ponds.

In macrophyte ponds the roots that are submerged will be the obvious sites for attachment by bacteria. The pathogen removal in waste stabilization ponds may include attachment of microorganisms to surface of macrophytes and algae as well as on the inner walls of the receptacles (containers or ponds) of the wastewater and suspended materials.

Studies conducted on attachment in pathogen removal in waste stabilization ponds and wetlands, focused on macrophytes because of the visible nature of their structure (Spira *et al.*, 1981; Kansiime and Nabulega, 1999). Algae being microscopic have not been considered as providing surface areas for attachment. In spite of several studies conducted in stabilization ponds (Pearson *et al.*, 1987; Curtis *et al.*, 1992; 1994; Pearson *et al.*, 1996; Davies-Colley *et al.*, 1999), the role of attachment as a pathogen removal mechanism using macrophytes and algae has not been adequately addressed.

The aim of this study is to determine the role attachment plays in faecal bacteria removal in macrophyte and algal ponds and more specifically on:

1.   The die-off rates of faecal bacteria in the presence and absence of macrophytes under batch-scale incubations;
2.   The die-off rates and mass balance of faecal bacteria on various surfaces in batch-scale incubations and
3.   The removal of faecal bacteria by attachment to harvested macrophytes in continuous flow treatment systems.

Materials and Methods

Batch-scale experiments

The removal of faecal bacteria in the presence and absence of macrophytes
Batch incubations studies were conducted to determine the removal rates of faecal bacteria in water lettuce (*Pistia stratiotes*), duckweed (*Lemna paucicostata*) and algal (seeded with algae and by natural colonization in raw wastewater) treatment systems. Each set up, except the seeded raw wastewater had six replicates. Three containers of the six replicates were kept in sunlight and the other three were kept in the dark. Domestic wastewater was obtained from the Kwame Nkrumah University of Science and Technology (KNUST) Kumasi, Ghana, domestic wastewater treatment plant. After thorough mixing to ensure an even distribution of bacteria, 800mL of sample was collected into 1L-plastic containers. The water lettuce-based system was mounted by using one water lettuce plant weighing 30g approximately with an average of 20 roots and large enough to cover the surface of the containers provided. The duckweed-based

system was set up by adding 7g fresh weight duckweed fronds to cover the surface (78.5cm^2) of plastic containers provided. The raw wastewater seeded with 20mL of water from an algal pond was also added. All sides of the containers were wrapped in aluminium foil to block sunlight penetration. The number of faecal bacteria suspended in the wastewater was measured every other day for 8 days at 8 GMT. The pH measurements were done with electronic meters (LF 323 — B/ SET 2, WTW – Germany). Faecal bacteria determinations were done using MacConkey broth after incubation at 44.5^{0}C for 24hrs with the Most Probable Number (MPN) technique. Green metallic sheen colonies were confirmed as faecal bacteria (Greenberg *et al.*, 1992).

Die-off and mass balance of faecal bacteria under batch-scale conditions
The die-off and mass balance of faecal bacteria attached to various parts of 1-L (Diameter of 11cm and of height 13cm) containers was carried out to determine the contribution each part exposed to the wastewater played in faecal bacteria removal. The number of faecal bacteria attached to various surfaces was carried out for duration of about one week. Faecal bacteria levels in suspension, attachment on the walls (surface area in contact with wastewater 440.7cm^2), sediments and on the roots and leaves of water lettuce and duckweed fronds (*Lemna paucicostata*) submerged in the wastewater were monitored every other day for 8 days under ambient conditions.

Faecal bacteria determinations were done using MacConkey broth after incubation at 44.5^{0}C for 24hrs with the Most Probable Number (MPN) technique. Green metallic sheen colonies were confirmed as faecal bacteria (Greenberg *et al.*, 1992). All faecal bacteria removal rates were calculated based on Chick's law. The percentage distributions of viable faecal bacteria on the various surfaces were also measured.

Wastewater characteristics
Characteristics of wastewaters used are presented in Table 1.

Table 1 Domestic wastewater characteristics

Parameter	Range	
	Kumasi (Ghana)	Ginebra (Colombia)
Temperature (^{0}C)	28-30	17-25
pH	6.8-8.8	7.0 - 7.2
DO (mg/L)	0-0.1	0.0-1.6
TDS (mg/L)	988-1112	742-938
TSS (mg/L)	112-140	12-225
BOD (mg/L)	260-320	284-360
COD (mg/L)	648-880	412-585
NH$_4$-N (mg/L)	32.4-48.8	19-52
NO$_2$-N (mg/L)	0.18-0.24	0-0.1
NO$_3$-N (mg/L)	2.6-2.9	0.0
Total phosphorous (mg/L)	3.2-5.0	4.3-9.4
Faecal bacteria (No/100mL)	2.1x 10^8- 3.7x10^8	2.4x10^7-1.8x10^7

Continuous flow systems

The experiment in the continuous flow system was carried out on bench and pilot scales in Ghana and Colombia respectively.

Bench-scale studies in Ghana

The number of faecal bacteria attached to water lettuce (*Pistia stratiotes*) and duckweed (*Spirodela polyhriza*) in a bench-scale continuous flow system in Ghana was measured by taking at random plants from each of the pond in series (Figure 1) during harvesting. The system consisted of 3 parallel treatment lines of 4 ponds each, involving the use of water lettuce (*Pistia stratiotes*), duckweed (*Spirodela polyrhiza*) and algae (natural colonization). A flow rate of $0.01m^3$/day was maintained in each treatment system. Each pond had a depth of 0.63m and surface area of $0.145m^2$ and hydraulic retention period of 7 days. Wastewater was collected at the influent grit chamber of KNUST wastewater treatment plant at 7 GMT daily and put into the anaerobic pond, which fed into the 3 pond systems, by gravity in a continuous flow (Figure 1). The water lettuce ponds were maintained by harvesting once every week. Duckweed ponds were harvested twice in a week.

Faecal bacteria attachment on water lettuce the roots were isolated by cutting roots from the main plant. The leaves were equally removed and faecal bacteria attachment on each part measured. All experiments were done in triplicates. The faecal bacteria populations attached to the macrophytes were determined using chromocult agar (Byamukama *et al.*, 2000) after grinding the plant biomass in a sterilized mortar. The percentage of faecal bacteria removed during harvesting by attachment was calculated based on macrophytes harvested in a week. The number of faecal bacteria attached to the various plants parts in the different ponds in series was almost the same whenever readings were taken. The ponds had actually stabilized and equilibrium had been reached. This study was just to measure the amount of viable faecal bacteria that was removed during harvesting of the macrophytes. The number of faecal bacteria removed by attachment on harvested macrophytes was measured in percentages.

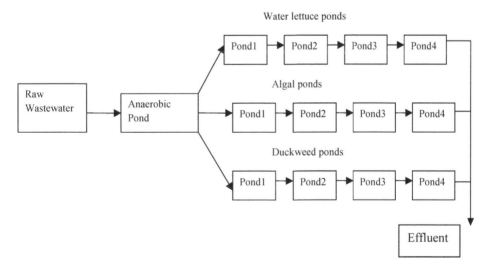

Figure 1 Schematic diagram of bench-scale macrophyte and algal waste stabilization ponds in Ghana

Pilot-scale

Pilot-scale experiments were conducted in Ginebra (Colombia, South America). A continuous flow system comprising of duckweed (*Spirodela polyrhiza*) and algal ponds was operated in parallel and received UASB pre-treated domestic wastewater from Ginebra Town (9000 inhabitants). A third pond also in parallel with the two other ponds was duckweed-based, but received raw wastewater without pre-treatment (Figure 2). Each pond was 64m in length and 0.7m in depth with a width of 4.95m for algal ponds and 5.2m for duckweed ponds. The L/W ratio was 12.9 for algal ponds and 12.3 for duckweed ponds. A flow rate of 16.6m^3/day was maintained in each pond. The theoretical retention period was 13.5 days. Beginning from the point of influent discharge, the ponds were marginally divided into three sections (A, B and C) with bamboo sticks on the surface to prevent the duckweed fronds from drifting (Figure 2). The study focused on faecal bacteria attachment in the two duckweed ponds. Duckweed coverage density of 600g/m^2 was maintained in both duckweed ponds and harvested twice in a week. Weekly harvested duckweed was weighed to determine the biomass yield per week in the duckweed ponds. Using the number of influent faecal bacteria discharged in a week and the number of viable faecal bacteria removed by attachment on the harvested macrophytes, the percentage of viable faecal bacteria removed by attachment on the harvested duckweeds was calculated. Faecal bacteria were determined by the Membrane Filtration technique in Membrane Lauryl Sulphate broth and incubated at 44.5^{0}C for 24hrs (Greenberg *et al.*, 1992).

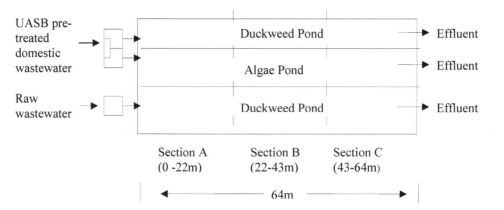

Figure 2 Schematic diagram of duckweed and algal ponds in a continuous flow system on a pilot scale in Ginebra Colombia

Results

Batch scale studies

The removal of faecal bacteria in the presence and absence of macrophytes
Algal colonisation was visible just after 3days of exposure of raw wastewater to sunlight. The removal rates of faecal bacteria in treatments under sunlight were higher than treatments under darkness.

Water lettuce had the lowest removal rates and the values were similar to those of raw wastewater under dark conditions. Faecal bacteria removal rates in the water lettuce containers exposed to sunlight were similar to those of algal-based treatment systems and raw wastewater with no differences. The removal rates in the algal treatment systems under sunlight exposures were higher than the others (Table 2).

Table 2 pH and faecal bacteria removal rates in macrophyte and algal ponds

Treatment system	pH	k (d^{-1})
Water lettuce in sunlight	6.7±0.3	2.4±0.9
Water lettuce in darkness	7.3±0.3	1.8±0.6
Algae seeded raw wastewater in sunlight	9.5±0.7	3.0±1.2
Raw wastewater in sunlight (Control)	9.3±1.0	2.8±1.1
Raw wastewater in darkness (Control)	7.3±1.0	2.0±0.2

Die-off and mass balance of faecal bacteria attachment under batch-scale conditions
It was observed that attachment of faecal bacteria occurred on all surfaces provided and continued to increase until the second day when decline in bacteria numbers were observed. Faecal bacteria attachments on the walls/cm^2 of wastewater containers with different plants were not significantly different on the 2nd and 4th day. Significant differences ($p < 0.05$) were observed on the 6th and 8th day between different treatments with the algal treatment system having the least faecal bacteria numbers/ cm^2 (Figure 3).

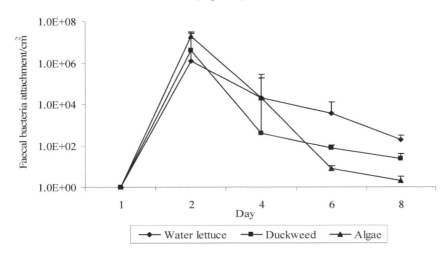

Figure 3 Attachment and removal of faecal bacteria to/from walls of macrophyte and algal ponds

The decline rates of faecal bacteria in suspension also followed a similar trend (Figure 4).

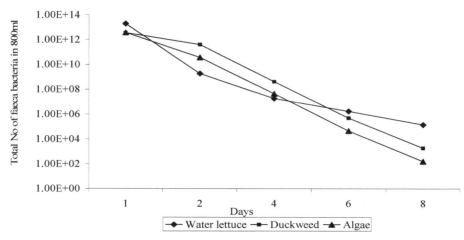

Figure 4 Removal rates of faecal bacteria in macrophyte and algal waste stabilization ponds

In comparison to the different treatments, the highest removal rate was observed in the suspension of the algae seeded treatment system (Table 3). The faecal bacteria removal rates were also higher in the algal treatment systems than macrophytes treatment systems (Table 3). Sediment die-off rates were highest in comparison with all other parts measured in all treatments. Within the water lettuce pond, the removal rate of faecal bacteria was highest on the leaves. Faecal bacteria removal rates on the walls of water lettuce containers were the lowest.

Table 3 Removal /die-off rates of faecal bacteria in various locations of macrophyte and algal ponds

Location	Pond system	$k\ (d^{-1})$
	Water lettuce	2.1±0.6
Wall	Raw wastewater seeded with algae	3.6±0.8
	Water lettuce	3.7±0.1
Sediment	Raw wastewater seeded with algae	4.3±0.1
	Water lettuce	2.4±0.2
	Water lettuce leaves	3.4±0.0
	Water lettuce roots	2.3±0.4
Suspension	Raw wastewater seeded with algae	4.3±1.8

Mass balance of faecal bacteria in various locations of macrophyte and algal ponds
The amount of faecal bacteria attached to the various locations was calculated as follows:
$$A_{roots} + A_{walls} + A_{sediment} + A_{suspension} + DB = IC$$
Where:
A_{roots} = amount attached to the water lettuce roots
A_{walls} = amount attached to the walls of the container
$A_{sediment}$ = amount attached to the sediments
$A_{suspension}$ = amount that remained in suspension
DB = unaccounted for due to death of faecal bacteria
IC =Initial concentration of viable bacteria (100%)

Table 4 Total numbers of viable faecal bacteria in various locations in water lettuce containers

Initial concentration of faecal coliforms in suspension in 800mL: $3.7 \times 10^{12} \pm 1.2 \times 10^{13}$					
Day	Leaves	Roots	Wall of container	Sediment	Suspension	Die-off fraction
1	0	0	0	0	3.7×10^{12}	0
2	4.8×10^6	5.7×10^{10}	1.3×10^{10}	7.2×10^8	4.2×10^{10}	3.6×10^{12}
4	6.6×10^5	4.0×10^9	1.4×10^8	7.2×10^7	3.3×10^8	3.7×10^{12}
6	3.5×10^4	2.4×10^7	5.7×10^6	4.5×10^4	2.9×10^6	3.7×10^{12}
8	3.9×10^3	4.5×10^2	4.8×10^3	24	1.7×10^5	3.7×10^{12}

In the water lettuce ponds most of the viable faecal bacteria were attached to the roots on the 2nd, 4th and 6th days. Most of the viable faecal bacteria were found in the suspension on the last day. Attachment of viable faecal bacteria in the sediments and on the roots and the walls of the container were low on the last day (Table 5). Log removal of faecal bacteria ranged from 3 on the leaves to 8 on the roots.

Table 5 Percentages of viable faecal bacteria in various locations in water lettuce containers

Day	Leaves (%)	Roots (%)	Wall of container (%)	Sediment (%)	Suspension (%)
1	0	0	0	0	100
2	0.0	50.6	11.5	0.6	37.3
4	0.0	88.1	3.1	1.6	7.3
6	0.1	73.4	17.4	0.1	8.9
8	2.2	0.3	2.7	0.0	94.9

In the algal ponds, the situation was not too different as high die-off of faecal bacteria occurred on all surfaces in the containers after attachment. The log removal ranged from 10 in the suspension to 7 on the walls (Table 6).

Table 6 Total numbers of faecal bacteria in various locations of algal ponds

Initial concentration of faecal bacteria in suspension in 800mL: $3.7 \times 1012 \pm 1.2 \times 10^{13}$				
Day	Wall	Sediment	Suspension	Die-off fraction
1	0	0	3.7×10^{12}	0
2	4.5×10^8	3.3×10^9	3.7×10^{10}	3.7×10^{12}
4	9.0×10^7	9.0×10^5	4.5×10^7	3.7×10^{12}
6	1.3×10^3	2.3×10^2	4.5×10^4	3.7×10^{12}
8	23	13	1.6×10^2	3.7×10^{12}

Most of the viable faecal bacteria were found in the suspension (Table 7). Most of the faecal bacteria were found in suspension on the 8th day (Table 6) but the levels were lower than that of the water lettuce ponds (Table 4-7).

Table 7 Percentages of viable faecal bacteria in various locations of algal ponds

Day	Wall (%)	Sediment (%)	Suspension (%)
1	0	0	100
2	1.1	8.1	90.8
4	66.2	0.7	33.1
6	2.8	0.5	96.7
8	11.7	4.7	81.6

Attachment in continuous flow systems

Bench-scale studies
The number of faecal bacteria in the ponds decreased in suspension along the series of ponds and likewise the number of faecal bacteria attached to macrophytes (Figure 5 and 6 respectively).

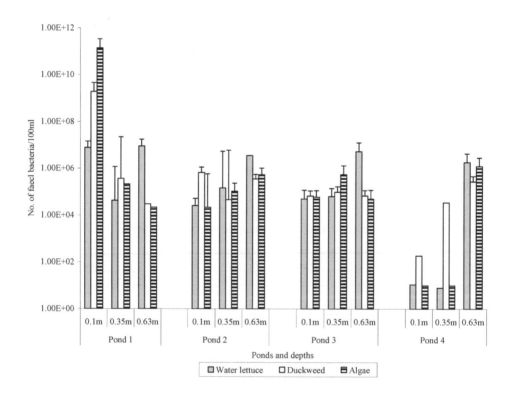

Figure 5 Faecal bacteria profile in bench-scale continuous flow macrophyte and algal ponds in Ghana

The number of faecal bacteria attached to the water lettuce plants was more than the number attached to the duckweed fronds per gram of fresh weight. The surface area available for attachment in the water lettuce ponds was much higher than that of the duckweed. The number of faecal bacteria removed through harvesting of the water lettuce and duckweed plants by attachment was less than 1% (Table 8).

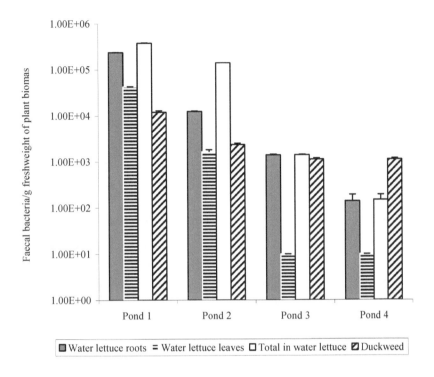

Figure 6 The number of faecal bacteria attached to macrophytes in a series of ponds in a bench-scale continuous flow system under equilibrium conditions in Ghana

Table 8 Faecal bacteria removal on harvested macrophytes in bench-scale continuous flow ponds in Ghana.

Water lettuce Treatment system	Faecal bacteria number inflow per week: $9.7 \times 10^{10} \pm 2.3 \times 10^{10}$	
	Harvested biomass (g/week)	Faecal bacteria removed weekly by attachment through harvesting
Pond 1	542±69	1.5×10^7
Pond 2	562±69	7.9×10^6
Pond 3	381±42	5.2×10^4
Pond 4	193±32	5.9×10^4
	Faecal bacteria removed by attachment to harvested water lettuce plants (%)/week: < 1%	
Duckweed treatment system	Faecal bacteria number inflow per week: $9.7 \times 10^{10} \pm 2.3 \times 10^{10}$	
	Harvested biomass (g/week)	Faecal bacteria removed weekly on harvested plant biomass
Pond 1	71±55	8.4×10^5
Pond 2	151±57	7.9×10^4
Pond 3	145±33	4.7×10^4
Pond 4	110±43	4.5×10^3
	Faecal bacteria removed by attachment to duckweed fronds harvested (%)/week: < 1%	

Pilot-scale studies

More faecal bacteria populations were found in the duckweed pond without treatment than the duckweed pond with the pre UASB treatment. Generally the last section C had the least faecal bacteria population at all depths in both treatment systems. For the surface portion where effluent discharge point was located the faecal bacteria population did not follow any trend (Figure 7). This suggests that series of ponds could be better treatment systems than single ponds.

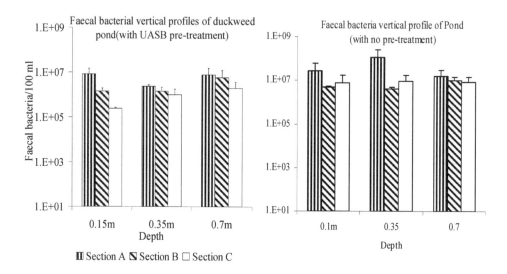

Figure 7 Faecal bacteria profile of duckweed ponds in Ginebra, Colombia

The number of faecal bacteria attachment in a single large pond in the Ginebra ponds did not follow any particular trend (Figure 8). More faecal bacteria were attached to duckweed in the non-UASB pre-treated pond than in the UASB pre-treated pond. The highest number of faecal bacteria attached to duckweed in the non-UASB pre-treated pond occurred in section C at the extreme end of the pond. In the UASB pre-treated pond the highest attachment of faecal bacteria to duckweed occurred in the middle section B but was not significantly different ($p < 0.05$) from the non-UASB pre-treated influent receiving pond. In sections A and C however, the results of the two treatments were significantly different ($p < 0.05$) (Figure 8).

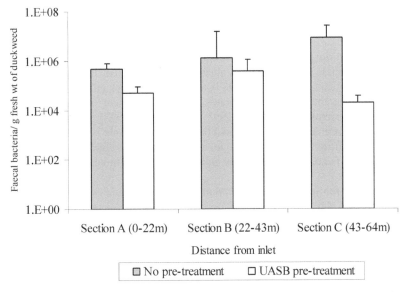

Figure 8 Attachment of faecal bacteria to duckweed fronds at various sections in UASB pre-treated and non pre-treated influent receiving duckweed ponds.

The amount of duckweed harvested in the UASB pre-treated influent receiving pond was higher than that of the non pre-treated influent receiving pond (Table 10). However, the contribution of faecal bacteria removal by attachment through harvesting was very low in both pond systems, i.e. less than 1% (Table 10).

Table 10 Faecal bacteria removal on harvested macrophytes in pilot-scale ponds in Ginebra, Colombia

Duckweed ponds (Non pre-treatment)	Faecal bacteria number inflow per week: $3.5 \times 10^{14} \pm 6.7 \times 10^{14}$	
	Harvested biomass (kg/week)	Faecal bacteria removed weekly on harvested plant biomass
Section A	56.0±13.4	2.8×10^7
Section B	59.3±10.6	5.9×10^7
Section C	57.1±23.7	4.6×10^8
	Faecal bacteria removed by attachment through harvesting /week: < 1%	
Duckweed ponds (UASB pre-treated)	Faecal bacteria number inflow per week: $1.3 \times 10^{14} \pm 2.3 \times 10^{14}$	
	Harvested biomass (kg/week)	Faecal bacteria removed weekly on harvested plant biomass
Section A	73.5±20.6	3.7×10^6
Section B	86.8±28.6	3.5×10^7
Section C	104.8±41.3	2.1×10^6
	Faecal bacteria removed by attachment through harvesting (%)/week: < 1%	

Discussion

Batch scale experiments

Removal of faecal bacteria in the presence and absence of macrophytes
Pathogen removal in macrophyte ponds has been attributed to attachment to vegetative roots (Kadlec and Hammer, 1982, Gersberg, *et al.*, 1987) and plant surfaces (Brix and Schierup, 1989). Other reports indicate that macrophyte ponds are more efficient in the removal of pathogens than algal pond systems (Mandi *et al.*, 1993; Garcia *et al.*, 1997). Our present study showed that in the presence of macrophytes the faecal bacteria removal rate is less than those in algal ponds though not significantly. Large surface area available in the algal ponds as well as high pH may be contributing factors operating in the algal systems to promote die-off.

Die-off and mass balance of faecal bacteria attachment under batch-scale conditions
The results of these studies showed that die-off occurs after faecal bacteria attachment to various surfaces. The fact that attachment of bacteria occurs on any available surface, both animate and inanimate objects was confirmed in this present study. Fletcher, (1996) reported that microbial cells attach firmly to almost any submerged surface in an aquatic environment. All root systems possess an anchorage for microorganisms around their thin film surfaces or rhizospheres (Kadlec and Hammer, 1982; Gersberg *et al.*, 1987).

The high faecal bacteria die-offs observed in the sediments (Table 4) also showed that when bacteria fall to the bottom they would probably die faster than when in suspension. The bottom layer tends to have high protozoan numbers, which could feed on the faecal bacteria Awuah *et al., unpublished*. After attachment and sedimentation, die-off begins and this contributed to the high faecal bacteria removal. The high numbers of faecal bacteria considered dead increased with time (Figure 3) showing that long retention periods are important contributing factors in faecal bacteria removal if conditions are unfavourable. In comparison to percentage of faecal bacteria attachment to surfaces with total viable bacteria numbers, attachment was substantially found to contribute to faecal bacteria removal. The die-off after attachment may be the reason for high faecal bacteria in macrophyte pond observed in chapter 7 of this thesis. The high removal rates of faecal bacteria observed in the algal colonised ponds could be due to high pH and DO fluctuations created in the algal ponds (Awuah *et al.*, 2001; 2004a; 2004b). The high pH could have had detrimental effect on the bacteria attached to the walls, sediments and in the suspension of the algal ponds thus, making the faecal bacteria removal faster at all locations in the algal ponds than those in the macrophyte ponds, which had neutral conditions. The algal system having high suspended solids provided a large surface area for attachment and coupled with the harsh environmental conditions had a higher die-off rate at all surfaces than macrophyte ponds. Attachment may thus be a very important mechanism in pathogen removal in algal ponds and must be investigated further. In the macrophyte pond where there were no suspended solids for attachment in the suspension phase high population of faecal bacteria remained in suspension even on the last day. Curtis *et al.*, (1992; 1994) claims that pH and DO are the sole mechanisms for faecal may not be completely true since attachment of the faecal bacteria may be of equal importance.

The suspended materials present in domestic wastewater have been proven as sites for faecal bacteria attachment (Fletcher, 1996; Kansiime and Nalubega, 1999; Donlan, 2002). Using the same concentration of faecal bacteria in the wastewater, the amount of faecal bacteria attached per gram was higher in the water lettuce ponds (Figure 4) than in the duckweed ponds. Water lettuce has more roots and root hairs, providing a much larger surface area for attachment than the duckweed, which had only two to three small roots per frond. When initial faecal bacteria concentrations were high, the amount of faecal bacteria attachment was also high.

Faecal bacteria attachment in continuous flow systems

In the bench-scale continuous flow systems, where the concentration of faecal bacteria was highest, a lot of detritus attachment was seen on the roots and they also had higher attachment of faecal bacteria on the macrophytes than the subsequent ponds (Figure 5). In the pilot-scale experiments, high faecal bacteria attachment was observed in the non-UASB pre-treated influent receiving pond than the UASB pre-treated ponds, which had lower faecal bacteria concentrations (Figure 8 and Table 10). The abundance of detritus in the water column increased the surface area for attachment of microorganisms (De Young, 1976; Dewedar and Beghat, 1995).

The removal of faecal bacteria by attachment through harvesting in the continuous flow experiments showed harvesting of macrophytes removes only a small percentage by attachment. This is because the treatment systems were already in equilibrium and die-off after attachment had taken place as illustrated in the batch-scale incubation studies. Bacteria die-off after attachment could not be measured. However; batch scale incubation studies have shown that die-off occurs after attachment and thus the more surface area available the more die-off that could occur to enhance pathogen removal. The number of faecal bacteria attachment in a single large pond in Ginebra, Colombia did not follow any particular trend. For aquaculture purposes and biomass resource recovery, ponds should be designed in series to ensure that the number of faecal bacteria attached to macrophytes is reduced at least in the last series of ponds as observed in the bench-scale continuous flow systems in Ghana.

The results of these batch and continuous flow systems experiments confirm several studies which also found microbial attachment to the roots of plants and to the suspended particles (Hughes and Rose (1971; Droopo and Ongley, 1974; Droopo and Jannasch, 1980; Klug and Reddy, 1984; Kansiime, and van Bruggen, 2001). Batch scale studies have also confirmed the importance of attachment in pathogen removal as stipulated by Gersberg *et al.*, (1978), Kadlec and Hammer, (1982); Ouazzani *et al.*, (1995) and Garcia *et al.*, (1997).

Algae can remain suspended in the wastewater and the removal of pathogens by attachment can occur when the algae with attached bacteria sink to the bottom because of increase in weight. Also, if the algae exert harsh conditions to the attached bacteria while they are in suspension, removal of bacteria can take place. The contribution of attachment to algae to faecal bacteria removal has not been investigated well enough (Fisher *et al.*, 1998) and more research needs to be done.

Conclusion

It may be deduced from these studies that faecal bacteria attach themselves to roots and leaf surfaces of water lettuce and duckweed fronts, suspended materials and other available surfaces in macrophyte and algal ponds. The attachment of faecal bacteria to macrophytes and other surfaces in macrophyte ponds may play a significant role for faecal bacteria removal through die-off. Large surface area availability may increase faecal bacteria attachment and subsequently increase the rate of its removal if conditions at attachment sites are unfavourable.

Under such conditions, long retention periods could be important in pathogen removal in both macrophyte and algal ponds.

Acknowledgement

This study was supported with grants from the Netherlands Government through the SAIL Foundation.

References

Atlas RM and Bartha R (1981). Microbial ecology; Fundamentals and applications. Addison – Wesley Publishers Co. London, 560p.

Awuah E, Anohene F, Asante K, Lubberding HJ and Gijzen HJ (2001). Environmental conditions and pathogen removal in macrophyte and algal-based domestic wastewater treatment systems. *Wat. Sci. Tech.* **44** (6), 11-18.

Awuah E, Kuffour AR, Lubberding HJ and Gijzen HJ (2002). Characterization and management of domestic wastewater in two suburbs of Kumasi. In: *Proceedings of Water and Health Conference* Ottawa, Canada, September 2002. 495p.

Awuah E, Lubberding HJ and Gijzen HJ (2004a). Effect of stable pH and pH fluctuations on faecal coliform removal in domestic wastewater. In: *KNUST SERR 3 Proceedings.* August 2003, Busua, Ghana, 300p.

Awuah E, Afealetse D-G, Lubberding HJ and Gijzen HJ (2004b). Comparative performance studies of macrophytes and algal-based in medium strength sewage. In: *KNUST SERR 3 Proceedings.* August 2003. Busua, Ghana 300p.

Bradling MG, Jass J and Lappin-Scott HM (1995). Dynamics of bacterial biofilm formation. In: Lappin-Scott HM, Costerton JW (eds.), *Microbial biofilms.* Cambridge University Press, pp. 46-63.

Byamukama D, Kansiime F, Mach RL and Farnleitner H (2000). Determination of *Escherichia coli* contamination with chromocult coliform agar showed a high level of discrimination efficiency for differing faecal pollution levels in tropical waters of Kampala, Uganda. *Appl. Environ. Microbiol.* **66**, 864-868.

Brix H and Schierup HH (1989). Use of aquatic macrophytes in water pollution control. *AMBIO*, **18** (2), 100-107.

Curtis TP, Mara DD and Silva SA (1992). Influence of pH, oxygen and humic substances on ability of sunlight to damage faecal coliforms in waste stabilization ponds. *Appl. Environ. Microbiol.* **58**, 1335-1343.

Curtis TP, Mara DD, Dixo NGH and Silva SA (1994). Light penetration in stabilization ponds. *Wat. Res.* **28** (5), 1031-11038.

Davies-Colley RJ, Donnison AM, Speed DJ, Ross CM and Nagels JW (1999). Inactivation of faecal indicator microorganisms in waste stabilization ponds interactions of environmental factors with sunlight. *Wat. Res.* **33** (5), 1220-1230.

Dewedar A and Bhgat M (1995). Fate of faecal coliform bacteria in a wastewater retention reservoir containing *Lemna gibba* L. *Wat. Res.* **29**, 2598 – 2600.

De Young J (1976). The purification of wastewater with the aid of rush or reed ponds. *Biolog. Cont. of Wat. Poll.* pp 133-139.

Donlan RM (2002). Biofilms: Microbial life on surfaces. *Emerg Inf. Dis.* **8**(9), 881-890

Droopo M and Ongley E (1974). Flocculation of suspended sediment in rivers of south eastern Canada, *Nat. Wat. Res. Inst.,* **28**, 1799-1809.

Droopo M and Jannasch H (1980). Advances of Aquatic Microbiology. Academic press, New York. pp. 23-35.

Fisher MM, Wilcox LW and Graham LE (1998). Molecular characterization of epiphytic bacterial communities on charophycean green algae. *Appl. Env. Microbiol.* **64** (11), 4384-4389.

Fletcher M (1996). Bacterial adhesion: Molecular and ecological diversity. John Wiley Inc. New York, 361p.

Garcia M, Bécarès E, Morris R, Grabow WOK and Joefre J. (1997). Bacteria removal in three pilot scale wastewater treatment systems for rural areas. *Wat. Sci. Tech.* **35** (11-12), 197-200.

Gersberg RM, Lyon SR, Brenner R and Elkins BV. (1987). Fate of viruses in artificial wetlands. *Appl. Environ. Microbiol.* **83**, 731-736.

Gray NF (2004). Biology of wastewater treatment. Second edition. Imperial College press London, 1421p.

Greenberg AE, Clesceri LS and Eaton AD (1992). Standard methods for examination of water and wastewater. 18th edition, American Public Health Association. American Water Works Association. Water Environmental Fed. Washington. D.C.

Hughes D and Rose A (1971). Microbes and biological productivity. Symposium 21 of the society for general microbiology held at University College London. Great Britain.

Kadlec R and Hammer D (1982). Pollutant transport in wetlands. *Environ. Progr.* **11**, 206-211.

Kansiime F and Nabulega M (1999). Wastewater treatment by natural wetland: the Nakivubo swamp, Uganda-Process and implications, PhD dissertation, UNESCO-IHE Delft, A.A Balkema Publishers, The Netherlands, 300p.

Kansiime F and van Bruggen JJA (2001). Distribution and retention of faecal coliforms in the Nakivubo wetland in Kampala, Uganda. *Wat. Sci. Tech.* **44** (11/12), 199-206.

Klug M and Reddy C (1984). Current perspectives in microbial ecology. In: *Proceedings of the Third International Symposium on Microbial Ecology* Michigan State University. August 1983. American Society for Microbiology. Washington. D.C. pp. 130-138.

Mandi L, Ouazzani N, Bouhoum K and Boussaid A (1993). Wastewater treatment by stabilization ponds with and without macrophytes under arid climate *Wat. Sci. Tech.* **28** (10), 177-181.

Mime Consult (2004). Final Draft reports Water and Sanitation policy for Community Water Supply and Sanitation Agency, June 38p.

Nhapi I, Dalu J, Ndamba J, Siebel MA and Gijzen HJ (2003). An evaluation of duckweed-based pond systems as an alternative option for decentralised treatment and re-use. *Wat, Sci. Tech.* **48**(2), 323-330.

Ouazzani N, Bouhoum K, Mandi L, Bouarab L, Habbari Kh, Rafiq F, Picot B, Bontoux, J and Schwartzbord J (1995). Wastewater treatment by stabilization pond: Marrakesh experiment. *Wat, Sci. Tech.* **31**(12), 75-80.

Pearson HW, Mara DD, Mills SW and Smallman DJ (1987) Physico-chemical parameters influencing faecal bacteria survival in waste stabilization ponds. *Wat. Sci. Tech.* **19** (12), 145-152.

Pearson HW, Mara DD, Cawley LR, Arridge HM and Silva SA (1996). The performance of an innovative tropical experimental waste stabilization pond system operating at high organic loadings. *Wat. Sci. Tech.* **33** (7), 63-73.

Spira WM, Hug A, Ahmed QS and Saeed YA (1981). Uptake of *Vibrio cholerae* Biotype *eltor* from contaminated by water hyacinth (*Eichhornia crassipes*). *Appl. Env. Microbiol.* **42,** 50-553.

Chapter Nine

Effect of Protozoa on Faecal Bacteria Removal in Macrophyte and Algal Waste Stabilization Ponds

Awuah E, Lubberding HJ and Gijzen HJ
Effect of protozoa on faecal bacteria removal in macrophyte and algal ponds

Effect of Protozoa on Faecal Bacteria Removal in Macrophyte and Algal Waste Stabilization Ponds

Abstract

Protozoa populations and the types of microbiota were determined at various depths (0.1m, 0.35m and 0.63m) in macrophyte and algal ponds receiving medium strength sewage. The set up consisted of 3 pond systems comprising water lettuce, duckweed and algal ponds operating in parallel. Each system consisted of four ponds operating in series with a hydraulic retention time of 7 days per pond. The effect of protozoa on the removal of faecal bacteria was determined through the elimination of protozoa by filtering wastewater from the second pond of each treatment system. In the water lettuce and duckweed ponds, protozoa were concentrated in the sediments. In the algal ponds protozoa were mostly found in the sediments and on the surface. The following protozoa were common in all the ponds: *Bodo, Vahlkampfia*. Other biota like copepods, mites and nematodes were also found in the 3 pond systems. The protozoan *Petalomonas* and *Chironomus* insect larvae were found only in the water lettuce and duckweed ponds. *Vorticella* and other unidentified ciliates were unique to the algal ponds. The algal ponds had the highest number of species diversity and the highest number of protozoa, followed by water lettuce and duckweed. The algal ponds were dominated by *Euglena*. Other algae such as *Chlorella, Chlorococcum, Phacus, Ulothrix,* and some diatoms were also found but in small quantities. When floating algae covered the ponds, *Spirogyra* and some coenocytic algae in strands dominated the algal community. The presence of protozoa on the removal of faecal bacteria was found to have a significant effect in the removal of *E. coli* and *Salmonella* ($p < 0.05$) in water lettuce pond system. In the duckweed and algal pond systems there was no difference between faecal bacteria removal in the presence and absence of protozoa. Predation by protozoa on faecal bacteria may be important in water lettuce ponds.

Keywords: Waste stabilization ponds, water lettuce; duckweed; algae; predation; protozoa, faecal bacteria

Introduction

Waste stabilization ponds (WSP), often referred to as oxidation ponds or lagoons, are holding basins used for wastewater treatment where decomposition of organic matter is processed naturally. The activity in WSP is a complex process involving bacteria, protozoa, algae, other microbes and metazoans, to stabilize the wastewater and to reduce pathogen populations. In the tropics where enteric diseases are common (Awuah *et al.*, 2002) the removal of pathogens is of much importance. The natural processes involved in pathogen removal in algal ponds, such as UV light, sunlight induced factors such as high pH and DO, have been studied in detail by several authors (Curtis *et al.*, 1992; Davies-Colley *et al.*, 1994; 1999) and have been found to be effective in pathogen removal. In macrophyte ponds these mechanisms are largely absent due to the shading provided by the macrophyte cover on the surface. Attachment to plant surfaces and predation of pathogens by protozoa are potential natural mechanisms, which could prevail in macrophyte ponds. Theoretically high pH and DO could play a role, because photosynthesis is still taking place, but previous studies have shown that these factors do no play an important role in macrophyte-based ponds (Awuah *et al.*, 2004). Predation of bacteria by protozoa has not been studied in detail in waste stabilization ponds, but this could provide an important mechanism for pathogen removal. Protozoa have been found to be effective in the removal of *Escherichia coli* (Curds and Fey, 1969). Other protozoa have been observed to feed on faecal coliforms, diphtherial, choleral, typhal and streptococcal bacteria (Enzinger and Cooper 1976; McCambridge and McMeekin, 1979). It has been reported that protozoa improve the effluent quality of activated sludge plants, trickling filters and rotating biological contactors (WPCF, 1990). Sinclair and Alexander (1989) also reported that slow growing bacteria are eliminated due to intense protozoa predation. Quantitative studies suggest that one *Tetrahymena pyriformis* cell is able to digest 500 bacteria per hour, which means that in 24hrs one individual protozoan can remove about 1.2×10^4 of bacteria. This suggests that predation alone can remove significant amounts of bacteria pathogens (Curds and Cockburn, 1968; 1971). In constructed wetlands predation of *Cryptosporidium* oocysts by ciliates such as *Euplotes patella* and *Paramecium caudatum* is reported as a mechanism for pathogen removal (Stott *et al.*, 2001).

The contribution of protozoa grazing on the removal of faecal bacterial in waste stabilization ponds has not been studied. Macrophyte ponds are now being used in developing countries for resource recovery, but their ability to remove pathogens and the mechanisms involved is still unknown (Awuah, 1999). An understanding of the role of bacteria predation by protozoa in the removal of faecal bacteria in macrophyte and algal pond systems may contribute to improved design and operation and maintenance practices of waste stabilization ponds for enhanced pathogen removal. The aim of this study was to determine the role of protozoa in the removal of faecal bacteria in macrophyte and algal ponds.

Materials and Methods

The experiments were carried out in a pilot-scale continuous flow system comprising of water lettuce *(Pistia stratiotes)*, duckweed *(Spirodela polyrhiza)* and algal (natural colonization) pond systems. An anaerobic pond with two days retention period preceded the set up. Each pond system consisted of four ponds operating in series and a retention period of 7 days in each pond with a total retention period of 28days. The flow rate in each treatment system was 0.01m^3 /day. The pond systems were operated in parallel according to the arrangement shown in Figure 1.

Water lettuce ponds were maintained by harvesting once every week and twice a week in the case of duckweed ponds.

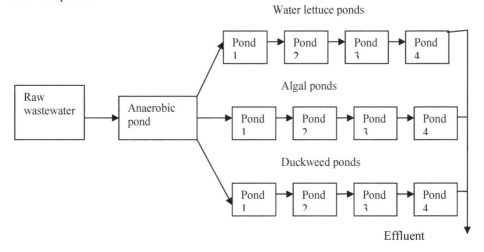

Figure 1 Schematic diagram of bench-scale macrophyte and algal waste stabilization ponds

Types of microbiota populations present and protozoan population profiles were studied in the pond systems. To determine the effect of protozoa on faecal bacteria in the pond systems, protozoa were removed through filtration of wastewater.

One litre of wastewater was collected from the surface of the 2nd pond of each pond system. Pond 2 was selected because this was the most productive pond in terms of macrophyte and algal growth. Five hundred ml of the 1L wastewater was filtered through a millipore nylon filter (NY41, pore size 7µm) to remove protozoa and algae. The filtered samples were examined to ensure that there were no protozoa and algae. The non-filtered samples contained protozoa and all other microbiota that were present. Non-filtered and filtered samples were divided into 3 equal volumes of 150ml in 3 plastic cups of 200mL capacity for each treatment. The set up was covered with aluminium foil to prevent the interference of light but loose enough to allow oxygen diffusion into the wastewater. The faecal bacteria populations were enumerated daily for 5 days using chromocult agar and incubated at 37^0C (Byamukama et al., 2000). E. coli was confirmed in EC medium with acid and gas production as positive. Coliforms were cultured on Endo agar as metallic sheen colonies. Salmonella was confirmed as growth in tetrathionate base broth (Greenberg et al., 1992). The removal rates in the filtered and non-filtered treatments were calculated using Chick's law.

In determining the protozoa populations and profiles, the number per mL were assessed at the surface (0.10m), middle portions (0.35m) and at the bottom (0.63m) in each month beginning in August 2001 to January 2002. A total of 50mL from 10 sampling points at each depth was collected and mixed by shaking manually. Samples were collected once in a month in the mornings between 6-8 GMT, counted and identified using a Sedgewick rafter counting chamber (Greenberg et al., 1992). Identification of protozoa was based on shape, size, morphology, type of motility and the presence of cilia and flagella in accordance with the guidelines provided by Finlay et al., (1988). A major limitation in this study was the fact that only about 50% of the protozoa could be identified.

Results

Effect of protozoa on faecal bacteria removal

Increase in bacteria numbers occurred in some of the treatments. In the water lettuce treatments increase occurred in faecal bacteria numbers in the absence of protozoa while decline in faecal numbers occurred in the non-filtered treatments in presence of protozoa. Differences were significant ($p<0.05$) for E. coli and Salmonella. There was no significant difference in die-off rates between filtered and non-filtered treatments for all faecal bacteria in duckweed and algal treatment systems and other coliform and other enterobacteria in the water lettuce treatment system (Table 1).

Table 1 Effect of the presence of protozoa on the removal of faecal bacteria

Treatment pond system	E. coli k (d^{-1})	Coliforms k (d^{-1})	Salmonella k (d^{-1})	Enterobacteria k (d^{-1})
Water lettuce non-filtered	0.2±0.0	1.1±0.1	0. 6±0.3	0.7±0.4
Water lettuce filtered	0.5±0.0*	1.1±0.9*	1.2±0.3*	0.3±1.1*
Duckweed non-filtered	2.8±2.4	1.2±1.1	0.1±0.0*	0.9±0.3*
Duckweed filtered	2.3±2.1	2.2±1.2*	0.6±0.0*	1.1±0.8*
Algae non-filtered	0.3±0.0	1.3±0.9	0.3±1.2	1.3±0.0
Algae filtered	0.5±0.2	1.7±1.4	2.9±2.4	0.8±0.4*

Increase in faecal bacteria numbers occurred all others values caused a decline in faecal bacteria numbers with time

Description of protozoa and other microscopic biota in the ponds

Several organisms were found in the wastewater treatment systems during the search for protozoa under the microscope. Protozoa of the genera *Bodo* and *Vahlkampfia* and algae of the genus *Euglena* were common to all the 3 pond systems. Other microbiota like mites, copepods and nematodes were also found in the 3 pond systems. The protozoa of the genus *Petalomonas*, and *Chironomus* insect larvae were found only in the water lettuce and duckweed ponds. *Vorticella* and other unidentified ciliates species were unique to the algal ponds. The algal ponds had the highest number of species diversity and the highest number of protozoa, followed by water lettuce and duckweed. *Euglena* dominated in all the algal ponds at most times. The following algae were also found in the algal ponds, although in small quantities: *Chlorella, Chlorococcum, Phacus, Ulothrix*, and some diatoms. When floating algae covered the ponds, *Spirogyra* and some coenocytic algae were dominant among the algal community (Table 2).

Table 2 Protozoa and other microscopic biota found in macrophyte and algal ponds

Treatment system	Amoebae	Ciliates	Flagellates	Algae	Metazoans
Raw sewage	-	2, 3, 4	1, 3	-	-
Anaerobic pond	2, 6, 8	2, 3, 4, 7, 10	3, 4	-	4
Water lettuce ponds					
Pond 1	3	2	1, 4	1, 6	1, 4
Pond 2	3, 4, 5	2	2	6	
Pond 3	3, 4, 5	1, 2, 6	2	*6*	
Pond 4	-	2, 9, 10	2	6	
Duckweed ponds					
Pond 1	-	2	1, 3	4, 6, 8	3
Pond 2	2, 7, 8	-	1	6	1
Pond 3	3, 8	-	1	*6*	2, 3
Pond 4	-	2	1, 4	6	1, 2, 3, 4
Algal ponds					
Pond 1	1, 2, 3,	3, 6, 7, 10	1	1, 2, 4, 6, 7, 8, 9	1, 3
Pond 2	2, 7, 8	2, 5, 8, 9	1	1, 2, 4, 5, 6, 7,8	1, 3, 4
Pond 3	2, 5	2, 5, 9	1	2, 3, 4, 5, 7, 8	1
Pond 4	2, 3, 4	2, 4, 5	1	1, 2, 4, 6, 7, 8	1, 3, 4

Key to Table 2

	Amoebae		Ciliates		Flagellates		Algae
1	(spike-like heliozoan)	1	(striped)	1	*Bodo*	1	*Chlorococcum*
2	*Vahlkampfia*	2	(hopping)	2	*Petalomonas*	2	*Chlorella*
3	(oval, large)	3	*Paramecium*	3	(spiral-shaped)	3	*Chlamydomonas*
4	*Platymoeba*	4	(double fan-shaped)	4	(larvae-like protozoa)	4	strands without cell walls
5	(small)	5	*Vorticella*			5	diatom*s*
6	(amoeba)	6	(pear-shaped)		**Metazoans**	6	*Euglena*
7	*Vanella*	7	*Tetrahymena*	1	copepods	7	*Phacus*
8	(large spherical)	8	(unidentified)	2	*Chironomus* larvae	8	*Spirogyra*
		9	*Stentor type 1*	3	mites	9	*Ulothrix*
		10	*Stentor type 2*	4	nematodes		

Protozoa population and profiles

The protozoa populations in the raw sewage were always small in quantity except in October 2001 and December 2001 when spiral shaped protozoa appeared in large quantities. The numbers declined drastically in January (2002) (Table 3).

A similar situation occurred in the 3 pond systems. Most of the protozoa were small in size (<20µm) and most of the small protozoa were flagellates. The medium sized protozoa (20-

50µm) were mostly ciliates and the large size protozoa (>50 µm) were mostly amoebae. The protozoa were not evenly distributed within the ponds systems.

Most of protozoa were found in the sediments in the macrophyte ponds. The lowest protozoa populations were found at the surface of the macrophytes ponds. Protozoa populations in the anaerobic pond were higher than in the raw wastewater. The algal ponds had the highest number of protozoa amongst the 3 pond systems. The protozoa populations in the water lettuce ponds were higher than that of the duckweed ponds. In the algal ponds high protozoa numbers were also observed on the surface as well as in the sediments. The duckweed ponds, which had the lowest protozoa population, showed complete absence at some depths (Table 6). Agglomeration of protozoa was also observed during the counting (Table 4-7).

Table 3 Size distribution of protozoa (per mL) in raw wastewater

Month	Size		
	<20µm	20-50µm	>50 µm
August 2001	1	12	4
September 2001	40	0	21
October 2001	3620	2	1
November 2001	1	3	7
December 2001	920	0	0
January 2002	1	2	0

Table 4 Size distribution of protozoa (no. per ml) in the anaerobic pond at various depths

Depth	Size		
	<20µm	20-50µm	>50 µm
0.10m	4227±2962	77±188	15±20
0.35m	1277±1267	115±128	16±13
0.63m	54±131	68±163	51±63

Table 5 Size distribution of protozoa (per mL) in water lettuce ponds at various depths

Pond	Depth	Size		
		<20µm	20-50µm	>50µm
1	0.10m	24±48	17±41	17±27
	0.35m	28±35	9±22	5±8
	0.63m	454±584	102±128	25±23
2	0.10m	12±26	2±2	7±12
	0.35m	61±150	54±102	9±11
	0.63m	1615±1151	405±465	52±98
3	0.10m	41±44	55±77	26±35
	0.35m	34±64	29±56	3±4
	0.63m	12488±20198	142±150	125±244
4	0.10m	28±40	29±53	79±179
	0.35m	176±167	39±22	44±44
	0.63m	5057±8330	254±279	35±38

Table 6 Size distribution of protozoa (per ml) in duckweed ponds at various depths

		Size		
Pond	Depth	<20μm	20-50μm	>50μm
1	0.10m	141±156	0±1	36±50
	0.35m	78±129	17±41	7±7
	0.63m	2143±4166	33±45	20±16
2	0.10m	239±448	52±122	2±4
	0.35m	84±176	2±4	1±1
	0.63m	1783±1548	80±100	30±17
3	0.10m	7±16	31±74	nil
	0.35m	12±18	118±268	1±1
	0.63m	590±395	845±1356	77±159
4	0.10m	3±8	nil	nil
	0.35m	518±1207	217±501	12±28
	0.63m	585±1012	228±478	nil

Table 7 Size distribution of protozoa (per ml) in algal ponds at various depths

		Size		
Pond	Depth	<20μm	20-50μm	>50μm
1	0.10m	1497±2277	209±448	68±80
	0.35m	874±1526	46±73	14±17
	0.63m	4218±9540	41±32	137±276
2	0.10m	2129±2563	73±152	51±79
	0.35m	373±818	14±24	6±7
	0.63m	6137±8459	340±448	185±256
3	0.10m	638±537	27±26	30±32
	0.35m	784±456	118±864	30±41
	0.63m	18083±23414	332±301	178±133
4	0.10m	6157±10302	165±318	100±159
	0.35m	951±921	106±149	40±48
	0.63m	9973±11241	371±316	108±168

Discussion

Effect of protozoa on faecal bacteria removal

Results from the water lettuce based-treatment systems showed that without protozoa higher faecal bacteria populations would be observed in effluents from waste stabilization ponds. Effective removal for faecal bacteria was observed in the water lettuce ponds in the presence of protozoa whiles faecal bacteria removal in the duckweed and algal ponds were not very different in the presence and absence of protozoa (Table 1). The differences in the contribution of protozoa towards faecal bacteria removal could be due to selective feeding by protozoa (Kinner *et al.*, 1998; Finley *et al.*, 1988). Selective feeding by protozoa is based on size, motility and the growth condition of the protozoa themselves. This could vary between pond systems and even from pond to pond in the same pond system. Differences in the number of protozoa and type of species in the 3 pond systems have been observed in this study (Tables 2-7). *Petalomonas* was present mostly in the water lettuce ponds (Table 2) and this flagellate might contribute to the removal of *E. coli* and *Salmonella* in this pond. The effect of protozoa grazing on the elimination of bacteria is well recorded. Predation by ciliates is considered to be a very important mechanism for bacteria removal in wastewater treatment plants (Curds and Vandyke, 1966; WPFC 1990; Curds, 1992; Decamp, *et al.*, 1999). Kinner *et al.*, (1998) reported that nanoflagellates could consume 12-74% of attached bacteria in a day. They considered protozoa predation as a major removal mechanism of pathogens in wastewater treatment systems.

There were no significant differences between the removal of faecal bacteria in duckweed and algal ponds in the absence and presence of protozoa. Although protozoa numbers were highest in algal ponds, the die-off rates were low. This means that high protozoa population does not always correlate with effective removal of faecal bacteria by predation. In the duckweed pond the limited effect of protozoa on bacterial removal could be due to both selective feeding and the low numbers of protozoa (Table 6). Curds and Vandyke, (1966) indicated that for protozoa to be effective in the removal of bacteria through predation, their number should be more than 1000/mL. Petropoulos *et al.*, (2003) reported that low protozoa numbers resulted in bacteria growth rate increases. This phenomenon is also very well known in rumen of cattle, where bacterial numbers may be doubled after removal of protozoa (Gijzen *et al.*, 1988). Increase in numbers of faecal bacteria numbers was observed in some of the treatments especially in the absence of protozoa. This increase in faecal bacteria numbers was also observed by Awuah *et al.*, (2004a) in batch incubation studies of domestic wastewater. Karpiscak *et al.*, (2000) reported an increase in total coliform numbers in the effluent of milk waste treated in constructed wetlands. Gibbs *et al.*, (1997) also reported of regrowth of faecal coliforms and *Salmonella* in biosolids.

These results show that the ability of protozoa to effectively remove faecal bacteria is not always clear in macrophytes and algal ponds. In the laboratory some positive results have been obtained by several authors (Curds and Vandyke, 1966; WPFC 1990; Curds, 1992; Decamp, *et al.*, 1999) but in the field the results could be different.

Faecal bacteria removal rates obtained here are comparable to what was observed in earlier pH effects studies. At extreme pH values of 4, 10 and 11 however, the removal rates obtained were higher than the results obtained during the present study of the same duration (Awuah *et al.*, 2004a). Awuah *et al.*, (2003) had shown that it was the high pH that was the major cause of faecal coliform removal and not the high dissolved oxygen.

Protozoa profiles and populations

In the 3 pond systems, most of the protozoa were found in the sediments. The high numbers of protozoa (>1000/mL) seen in all 3 pond systems suggests that predation of protozoa on faecal bacteria could be important in the sediments. Gannon *et al.*, (1983) and Awuah *et al.*, (2004a), have shown that number of bacteria in the sediments were always more than at the other depths. This suggests that high numbers of protozoa cannot be correlated with effective removal faecal bacteria as claimed by Curds and Vandyke, 1966 and Decamp *et al.*, 1999 at all times.

The profile of the protozoa shows that protozoa populations are not evenly distributed in the pond systems (Table 5-7). Hence, in areas where there are no protozoa, effective removal of faecal bacteria through protozoa predation of faecal bacteria would not occur. The low protozoa population especially at the surface of the macrophyte ponds shows that complete reliance on predation for faecal bacteria removal may not be adequate.

Conclusion

It may be concluded that protozoa may play a major role in the removal of *E. coli* and *Salmonella* in water lettuce ponds. However, the removal of faecal bacteria will not be significant in duckweed and algal ponds especially when the protozoa numbers are low. Long retention periods in the presence of harsh environmental conditions may be the main contributing factors in the removal of faecal bacteria in macrophyte and algal-based waste stabilization ponds. Design of macrophyte ponds therefore should consider promoting harsh environmental conditions and increase in depth for longer retention periods in the ponds.

Acknowledgement

The authors wish to express their appreciation to SAIL Foundation for sponsoring this research and Mr. Ofori Donkor of KNUST wastewater treatment plant who fed the 3 pond systems with domestic wastewater on a daily basis.

References

Athayde STS, Pearson HW, Silva SA, Mara DD, Athayde Jnr. GB and de Oliveira R. (2000). Algological study in waste stabilization pond. Conferencia Latinoamericano, Lagunas de estabiliztaion y reuso. Cali, Colombia, pp. 132-139.

Awuah E, Kuffour AR, Lubberding HJ and Gijzen HJ (2002). Characterization and management of domestic wastewater in two suburbs of Kumasi. In: *Proceedings of Water and Health Conference*, Ottawa, Canada, September 2002.475p.

Awuah E, Boateng J, Lubberding HJ and Gijzen HJ (2003). Environmental conditions and effect of pH on faecal coliforms in domestic wastewater: In: *Proceedings of KNUST SERR 2*. Elmina, Ghana. August, 2002. 261p

Awuah E, Lubberding HJ and Gijzen HJ (2004a). The effect of stable pH and pH fluctuations on faecal bacteria in domestic wastewater. In: *KNUST SERR 3, Proceedings*. Busua, Ghana. September 2003 300p.

Awuah E, Oppong-Peprah M, Lubberding HJ and Gijzen HJ (2004b). Comparative performance studies of macrophyte and algal-based stabilization ponds. *J. Toxicol. and Environ. Health*. Part A, **67**, 1-13.

Byamukama D, Kansiime F, Mach RL and Farnleitner H (2000). Determination of *Escherichia coli* contamination with chromocult coliform agar showed a high level of discrimination efficiency for differing faecal pollution levels in tropical waters of Kampala, Uganda. *Appl. Environ. Microbiol.* **66**, 864-868.

Curds CR and Vandyke JM (1966). The feeding habits and growth rates of some fresh ciliates found in activated sludge plants. *J. Appl. Ecol.* **3**, 127-137.

Curds CR and Cockburn A (1968). Studies on growth and feeding of *Tetrahymena pyriformis* in axenic and monoxenic culture. *J. Gen. Microbiol.* **54**, 343-358.

Curds CR and Fey GJ (1969). The effect of ciliated protozoa on the fate of *Escherichia coli* in the activated sludge process. *Wat. Res.* **3**, 853-867.

Curds CR and Cockburn A (1971). Continuous monoculture of *Tetrahymena pyriformis*. *J. Gen. Microbiol.* **66**, 95-102.

Curds CR (1992). Protozoa in the Water Industry. Cambridge University Press, Cambridge, pp. 122.

Curtis TP, Mara DD and Silva SA (1992). Influence of pH, oxygen, and humic substances on ability of sunlight to damage faecal coliforms in waste stabilization pond water. *Appl. Environ. Microbiol.* **58**, 1335-1343.

Davies-Colley RJ, Bell RG and Donnison AM (1994). Sunlight wavelengths inactivating faecal indicator micro-organisms in waste stabilization ponds. *Wat. Sci. Tech.* **35**(11/12), 219-225.

Davies-Colley RJ, Donnison AM, Speed DJ, Ross CM and Nagels JW (1999). Inactivation of faecal indicator microorganisms in waste stabilization ponds: Interactions of environmental factors and sunlight. *Wat. Res.* **33,** 1220-1230.

Decamp O, Warren A and Sanchez R (1999). The role of ciliated protozoa in subsurface flow wetlands and their potential as bio-indicators. *Wat. Sci. Tech.* **40**(3), 91-98.

Enzinger RM and Cooper RC (1976). Role of bacteria and protozoa in the removal of *Escherichia coli* from estuarine waters. *Appl. Environ. Microbiol.* **31**, 758-763.

Finlay BJ, Rogerson A and Cowling AJ (1988). A beginners' guide to the collection, isolation cultivation and identification of freshwater protozoa. Natural Environmental Research Council, UK, pp. 78.

Gijzen HJ, Lubberding HJ, Gerhardus M and Vogels GD (1988). Contribution of rumen protozoa to fibre degradation and cellulase activity in vitro. *FEMS Microbiology Ecology*, **53**: 35-44*).*

Gibbs RA, Hu CJ, Ho GE and Unkovich I (1997). Regrowth of faecal coliforms and *Salmonellae* in stored biosolids and soil amended with biosolids. *Wat. Sci. Tech.* **35**(11/12), 269-275.

Greenberg AE, Clesceri LS and Eaton AD (1992). Standard methods for examination of water and wastewater. 18th edition, American Public Health Association. American Water Works Association. Water Environmental Fed. Washington. D.C.

Karpiscak MM, Sanchez LR, Freitas RJ and Gerba CP (2000). Removal of faecal bacteria indicators and pathogens from dairy wastewater by a multi-component treatment system. *Wat. Sci. Tech.* **44** (11/12), 83-190.

Kinner NE, Harvey RW, Blackeslee K, Navarino G and Meeker A (1998). Size selective predation on ground water bacteria by nanoflagelates in organic contaminated aquifer. *J. Environ. Microbiol.* **64**, 818-62

McCambridge J and McMeekin T (1979). Protozoan predation of *Escherichia coli* in estuarine waters. *Wat. Res.* **13**, 659-663.

Palmer CM (1969). A composite rating of algae tolerating organic pollution. *J. Phycol.* **5,** 78-82.

Petropoulos P, Wright DA and Gilbride K (2003). The effect of protozoa grazing on nitrification in activated sludge http://www.csmscm.org/english/abstracts/public/view_abs.asp?id=453.

Sinclair JL and Alexander M (1989). Effect of protozoa predation on relative abundance of fast and slow growing bacteria. *Can. J. Microbiol.* **35**, 578-582.

Stott R, May E, Matsushita E and Warren A (2001). Protozoa predation as a mechanism for the removal of *Cryptosporidium* oocysts from wastewaters in constructed wetlands. *Wat. Sci. Tech.* **44**(11-12), 191-198.

Vymazal J, Sladecek V and Stach J (2001). Biota participating in wastewater treatment in horizontal flow constructed wetland. *Wat. Sci. Tech.* **44**(11/12), 211-214.

WPCF (1990). Wastewater biology: The microlife special publication by Water Pollution Control Federation. Alexandria Virginia, 196p.

Summary in English

The pathogen removal mechanisms in macrophyte and algal ponds were studied in Ghana and Colombia. The macrophytes used were water lettuce (*Pistia stratiotes*) and duckweeds (*Lemna paucicostata* and *Spirodela polyrhiza*). The selection of the species was based on economic importance and availability. *Lemna* was used during the first investigations and when it was realized it could not withstand the ammonia levels in the wastewater used, *Spirodela polyrhiza* a more rare species in Ghana was used. The main mechanisms considered in this study were pH, protozoa predation and attachment. The microorganisms used were faecal bacteria namely, *Escherichia coli*, coliforms, *Salmonella sp.* and other enterobacteria isolated on chromocult agar and *E. coli* ATCC13706 and enterococci isolated on Slanetz Bartley medium. Studies were conducted under batch scale and continuous flow systems using bench scale and pilot scale ponds.

Chapter 1
Under this chapter a review of existing wastewater management systems in Ghana and the general pathogen removal mechanisms in wastewater stabilization ponds were evaluated. The aim and objectives of the PhD research were also outlined.

Chapter 2
Batch scale performance studies of stabilization ponds with and without macrophytes were carried out. Results showed that there were no significant differences in the treatment efficiencies of the domestic wastewater between macrophyte and algal ponds. Enterococci bacteria removals in all the treatment systems were similar even under dark conditions. This showed that mechanisms other than sunlight-induced conditions in pond systems might be responsible for pathogen removal. The interpretation of these results must be done with caution since the indicator organism used in this study are faecal enterococci and does not represent all groups of pathogens and also for the fact that the experiment was done on a batch scale.

Chapter 3
In assessing the environmental conditions in a continuous flow systems and the effect of pH on enterococci removal in the macrophyte and algal ponds, low pH were found to be more bactericidal on enterococci than alkaline conditions in domestic wastewater. The bactericidal properties at different pH values were enhanced by sunlight. It was also observed that light, nutrient depletion, low pH and the long retention period in the continuous flow systems might have played a key role a in the removal of enterococci in the macrophyte and algal ponds. Pathogen removal mechanisms in macrophyte and algal ponds could be different. The effect of pH on faecal coliforms and other pathogen indicators, under similar conditions should be studied for comparison.

Chapter 4
The effect of pH on faecal coliforms was done after intensive measurements of environmental conditions in the continuous flow system. It was concluded from this study that sunlight enhanced faecal coliform removal at pH 4, 7 and 9 but not at pH 5 and 10. Also, pH 5 was least detrimental to faecal coliforms. High pH was more effective in the removal of faecal coliforms than DO.

Chapter 5
Effect of pH fluctuations on different indicator organisms was assessed. The results showed that under different pH treatments faecal bacteria behave differently. This study shows that pH

fluctuations, extremes of pH, presence of other microorganisms may all contribute to die-off in stabilization ponds. Fluctuation from 7-9 common in algal ponds may contribute to *E. coli* elimination in algal ponds.

Chapter 6
An assessment of comparative performance studies was conducted using low sewage strength. Algal ponds were found to be more efficient in the removal of pathogens than macrophytes, while macrophytes were more efficient in organic load and nutrient removal than algal ponds. Faecal coliform profiles showed that in the water lettuce ponds if the effluent discharge had been located n the middle portion the effluent quality would have been the same as that of the algal pond system. Sludge accumulations in the macrophyte ponds were found to be several times lower than algal pond systems. Macrophytes may decrease the frequency of desludging through low sludge accumulations. Duckweed cover prevented the breeding of mosquitoes in the ponds. The benefit of macrophyte and algal waste stabilization ponds could be harnessed by combining both systems. Since this study was conducted by diluting sewage, it was recommended that performance studies be repeated using raw wastewater, which is of medium strength for confirmation.

Chapter 7
Using the medium strengths sewage comparative performances studies were conducted for a period of one year. Macrophytes performed equally well in the removal of organic load, nutrients and faecal bacteria removal if the pond systems are well managed. The mechanisms in the removal of faecal bacteria were attributed mainly to attachment, sedimentation, predation, nutrient depletion and the long retention periods in all pond systems. In the algal ponds pH and DO concentrations were suggested as the main cause of pathogen removal. The macrophyte waste stabilization ponds however, cannot be used for the treatment of domestic wastewater if the culture of maintenance is very poor in a given community. From the high biomass yields associated with the macrophytes, if they are not harvested, the ponds may be eutrophied in a very short time and create a major environmental nuisance.

Chapter 8
The importance of attachment in pathogen removal was assessed in this study. The results showed that faecal bacteria attach to walls of containers holding wastewater, water lettuce roots and leaves, duckweed fronds and algae. When the die-off rates and mass balance of faecal bacteria on various surfaces in batch-scale incubations were studied, die-off was observed immediately after attachment. Higher die-off was observed in surfaces in the algal ponds. Most of the viable bacteria were found attached to water lettuce roots and to suspended solids in the algal containers. Harvesting of macrophytes removed <1% of viable faecal bacteria in continuous flow ponds in Ghana and in Colombia. In comparison to percentage of faecal bacteria attachment to surfaces with total viable bacteria numbers, attachment was substantially found to contribute to faecal bacteria removal (>70). Attachment and subsequent settling of suspended solids contributes to prolonged retention of faecal bacteria in stabilization ponds, and as such provides the conditions for die-off.

Chapter 9
The effect of the presence of protozoa on faecal bacteria removal was measured. The types of protozoa seen in the ponds were counted and identified. Algal ponds had the highest species diversity and highest concentration of protozoa. It was concluded that protozoa might play a major role in pathogen removal in macrophyte and algal ponds. Predation may suppress microbial populations in waste stabilization ponds.

Concluding remarks

Based on the results of all the mechanisms studied so far, a relative importance of all the mechanisms of importance in macrophyte and algal pond systems are presented in Table 1. In order of importance, long retention periods, attachment, sedimentation, predation and low pH are mechanisms in macrophyte ponds enhancing faecal bacteria removal. The low pH is not given a high priority because the conditions must be created in low strength sewage, a practice that could be rare in developing countries with water scarcity problems. Also, low pH was only found in water lettuce ponds and was more applicable to enterococci bacteria. In the algal ponds, it is long retention periods, sunlight penetration, attachment, high pH, pH fluctuations and sedimentation, which are the mechanisms of importance in faecal bacteria removal. Presence of protozoa was also found to be important but true grazing studies could not be quantified.

Table 1 Relative importance of mechanism in faecal bacteria removal in macrophyte and algal ponds studied

Mechanism	Macrophyte pond systems	Algal pond system
Sedimentation	++++	+++
Attachment	++++	++++
DO concentrations	+	+
High pH>9	0	++++
Low pH<5	++	0
pH fluctuations	0	+++
Sunlight penetration	0	+++
Temperature fluctuations	0	0
Presence of protozoa	+++	+++
Retention period	++++	++++
Natural die-off	++++	++++

+ Degree of importance, 0 Not important

Based on our experiments DO did not play a major role in faecal bacteria removal. Retention period emerges as the most important factor in macrophyte ponds and designs of ponds should be deeper than the current depth of 0.5-1.0m to save space and improve on performance.

Summary in Dutch

Dit proefschrift gaat over de verwijdering van pathogenen uit afvalwateroxidatievijvers in Ghana, gedomineerd door watersla (*Pistia stratiotes*) en eendenkroos (*Lemna paucicostata*, *Spirodela polyrhiza*) of door algen. Begonnen werd met *Lemna*, maar toen bleek dat *Lemna* niet tegen de hoge ammoniumconcentraties in het afvalwater kon, is het onderzoek vervolgd met *Spirodela*, een veel zeldzamere soort in Ghana. De bestudeerde microörganismen waren faecale bacteriën: *Escherichia coli*, coliformen, *Salmonella sp.*, enterobacteriën en enterococcen. De resultaten zijn niet alleen afkomstig van experimenten op laboratoriumschaal in Ghana, maar ook van "pilot scale" vijvers in Colombia.

Hoofdstuk 1
In hoofdstuk 1 wordt de stand van zaken met betrekking tot afvalwaterzuivering in Ghana gegeven. Ook komen de mechanismen van pathogenenverwijdering in oxidatievijvers aan de orde. Tenslotte worden de doelstellingen van dit onderzoek geformuleerd.

Hoofdstuk 2
In hoofdstuk 2 worden oxidatievijvers met en zonder macrofieten (dus met algen) vergeleken. Geen significante verschillen werden gevonden in zuiveringsefficiëntie van huishoudelijk afvalwater tussen de twee systemen. Het feit dat enterococci in beide systemen even snel werden verwijderd in licht en in donker houdt in dat zonlicht geen essentiële rol speelt. Competitie met heterotrofe bacteriën om nutriënten zou in dit geval de verdwijning van pathogenen kunnen verklaren. Omdat deze experimenten op "batch scale" zijn uitgevoerd en gewerkt is met enterococci als indicatororganisme is enige voorzichtigheid geboden bij de interpretatie van de resultaten.

Hoofdstuk 3
De pH lijkt een belangrijke rol te spelen bij de verwijdering van enterococci in macrofieten- en algenvijvers: zure omstandigheden zijn werkzamer dan basische. Zonlicht heeft een bijkomend effect, zuurstof lijkt nauwelijks een rol te spelen. Enterococci kunnen goed overleven bij hoge pH, speciaal in het donker. De omstandigheden zijn nogal verschillend in de oxidatievijvers: in waterslavijvers is het water zuur, in eendenkroosvijvers neutraal en in algenvijvers basisch. De combinatie van zuurstof, licht, lage pH, tekort aan nutriënten en een lange verblijftijd in de "continuous flow" experimenten lijken alle bij te dragen aan de verwijdering van de enterococci.

Hoofdstuk 4
Door middel van een intensief meetprogramma is de invloed van pH op faecale coliformen bestudeerd. Zonlicht stimuleerde hun verwijdering tussen pH 4 en 9, maar niet bij pH 10; een verklaring voor het afwijkende resultaat bij pH 5 kon niet gegeven worden. Verhoogde zuurstofconcentraties hadden minder effect op de verwijdering van faecale coliformen dan verhoogde pH.

Hoofdstuk 5
In hoofdstuk 5 wordt het effect van pH veranderingen op verschillende indicatororganismen vergeleken. Extreem hoge of lage pH waarden bespoedigen de verwijdering van *E.coli* en coliformen, terwijl pH fluctuaties belangrijker zijn voor de verwijdering van *Salmonella*, enterobacteriën en enterococci. Zure omstandigheden bespoedigen de verdwijning van

enterococcen, maar hebben juist minder effect op *E.coli*, *Salmonella* en coliformen. Extreme pH, pH fluctuaties en de aanwezigheid van andere micro-organismen kan allemaal bijdragen aan de verwijdering van pathogenen in oxidatievijvers.

Hoofdstuk 6

In hoofdstuk 6 worden macrofietenvijvers met algenvijvers vergeleken in hun algemene vermogen tot zuivering van laagbelast afvalwater. Algenvijvers zijn beter in het verwijderen van pathogenen, macrofietenvijvers in de verwijdering van organische stof en nutriënten. De ophoping van slib in algenvijvers is veel groter, zodat uit macrofietenvijvers veel minder vaak slib hoeft te worden verwijderd. Het dek van eendenkroos verhinderde muskieten om eieren in het water af te zetten. De voordelen van beide systemen kunnen worden benut door ze te combineren. Het onderzoek zou herhaald moeten worden voor echt, sterker geconcentreerd, afvalwater.

Hoofdstuk 7

Als macrofietenvijvers goed worden onderhouden presteren ze even goed als algenvijvers in de verwijdering van organische stof, nutriënten en faecale bacteriën uit geconcentreerd afvalwater. De verwijdering van faecale bacteriën uit de macrofietenvijvers werd toegeschreven aan hechting, sedimentatie, predatie, nutriëntentekort en de lange verblijftijd, terwijl DO en de pH de belangrijkste factoren waren in algenvijvers.

Macrofieten verwijderen fosfor slechts in beperkte mate; wetgeving en politieke acties zijn daarom nodig om het fosforgehalte in huishoudelijk afvalwater omlaag te brengen. Afvalwateroxidatievijvers zijn goed toepasbaar in Ghana, maar voordat macrofieten worden gebruikt moet er een goede voorlichtingscampagne worden opgezet. Slecht onderhouden macrofietenvijvers hebben namelijk een negatief effect, mede door een snelle eutrofiëring van het water.

Hoofdstuk 8

In hoofdstuk 8 is het belang van hechting onderzocht voor de verwijdering van faecale bacteriën.

Zowel uit de "batch scale" als de "continuous flow" experimenten blijkt dat faecale bacteriën aan alle beschikbare oppervlakken hechten: aan wortels en bladeren van watersla en eendenkroos, aan algen en wat er ook maar beschikbaar is in algen- en macrofietenvijvers. Dus hoe meer oppervlak er beschikbaar is, hoe meer faecale bacteriën er hechten en hoe meer er verwijderd zullen worden. Ook door een langere retentietijd kunnen er meer hechten. Het oogsten van de macrofieten heeft geen merkbaar effect op de verwijdering van faecale bacteriën.

Hoofdstuk 9

De invloed van predatie door protozoën is bestudeerd in hoofdstuk 9. Daarvoor werden de protozoën geïdentificeerd en geteld. Algenvijvers hadden de grootste diversiteit en de hoogste aantallen. Protozoën zouden wel eens een belangrijke rol kunnen spelen bij de verwijdering van pathogenen in algen- en macrofietenvijvers, maar zouden ook de gehele bacteriepopulatie negatief kunnen beïnvloeden.

Conclusies

Op grond van de resultaten in dit proefschrift is in tabel 1 een overzicht gegeven van de verschillende verwijderingsmechanismen voor faecale bacteriën en hun relatieve belangrijkheid.

Tabel 1 Relatieve bijdrage van het mechanisme voor het verwijderen van faecale bacteriën uit macrofieten- en algenvijvers.

Mechanisme	Macrofietenvijvers	Algenvijvers
Sedimentatie	++++	+++
Hechting	++++	++++
O_2 concentratie	+	+
Hoge pH (>9)	0	++++
Lage pH (<5)	++	0
pH fluctuaties	0	+++
Zonlicht	0	+++
Temperatuur variatie	0	0
Aanwezigheid van protozoën	+++	+++
Retentietijd	++++	++++
Natuurlijk afsterven	++++	++++

+: *graad van belangrijkheid,* 0: *niet relevant*

De volgorde van relevante mechanismen voor het verwijderen van faecale bacteriën uit macrofietenvijvers is: retentietijd, hechting, sedimentatie, predatie en lage pH; uit algenvijvers: retentietijd, zonlicht, hoge pH, hechting, sedimentatie. Aanwezigheid van protozoën speelde ook een belangrijke rol, maar hun bijdrage aan de verwijdering kon niet gekwantificeerd worden. Zuurstof daarentegen leek niet belangrijk te zijn.

Samenvattend kan gesteld worden dat de twee belangrijkste factoren retentietijd (met als gevolg natuurlijke afsterving) en hechting (met als gevolg sedimentatie) zijn. Om de retentietijd te verlengen, en dus de kans op natuurlijke afsterving en sedimentatie te kunnen verhogen, zouden vijvers dieper gemaakt kunnen worden dan de huidige 0.5 tot 1 meter.

Curriculum Vitae

Esi Awuah was born on the 14th of September 1954 at Akim-Oda in the Eastern Region of Ghana. She attended Anglican Primary and Antwi Banson Middle School all at Akim-Oda. In 1968 she entered the Akim Oda Secondary School and graduated in 1973. She entered Aburi Girls Secondary School that same year and graduated in 1975. She was enrolled in the Biological Science degree program at the Kwame Nkrumah University of Science and Technology where she obtained a Bachelor of Science Honours degree with a major in microbiology.

In 1981, she received a WHO award in connection with the Water and Sanitation Decade to do her masters in Environmental Science at the College of Environmental Science and Forestry (SUNY) Syracuse, New York, USA. She obtained her Master of Science in 1985. Her thesis was on "Mesophillic composting of sewage sludge, corncobs and sawdust and its effect on pathogen removal". She was appointed as a lecturer in 1986 and promoted to senior lecturer in 1996. In 2002, she was promoted to the position of Associate Professor in 2002. In 1998, she obtained a fellowship from the Netherlands Government to pursue a PhD degree within the Water and Sanitation Sector Capacity building program in the Department of Civil Engineering, KNUST Kumasi, Ghana.

She is married to Prof. Richard Tuyee Awuah a professor in plant pathology and Dean Faculty of Agriculture KNUST, Kumasi, Ghana. In addition her marriage has been blessed with two handsome boys Ato and Kobbina who are respectively studying Economics and Law at KNUST Kumasi, Ghana and Aerospace Engineering/Pre-med at Binghamton University, USA.
She is a devout Christian, a church organist, a choir mistress, a Bible class leader and a Local Preacher in the Methodist Church of Ghana.

She is currently an Associate Professor and the Sectional Head of Environmental Quality Engineering Division in the Department of Civil Engineering at the Kwame Nkrumah University of Science and Technology, Kumasi, Ghana
Her email address is **esiawuahrt@yahoo.com**
Her telephone numbers are

1.	233 (0) 51 60235 (O).
2.	233 (0) 51 60192 (H)
3.	233 (0) 244 786791 (Mobile)

The research and the whole PhD program was financed by the Dutch Government through the collaboration project in the Water and Sanitation Sector (WSESP) capacity building between KNUST and UNESCO-IHE Institute for Water Education. She has also personally spent quite some amount of money on this project.